EVOLUTIONARY ESSAYS

Kyle Lance Proudfoot

authorHOUSE®

AuthorHouse™ UK Ltd.
1663 Liberty Drive
Bloomington, IN 47403 USA
www.authorhouse.co.uk
Phone: 0800.197.4150

Published by AuthorHouse 05/19/2014

ISBN: 978-1-4969-8082-3 (sc)
ISBN: 978-1-4969-8083-0 (e)

Prologue

Equal In Return
(The Fact You Lack It You Are Attracted)
Written by: Kyle Lance Proudfoot ©®™
Published by: AuthorHouse UK

Introduction

This is everything I see as positive for this Society, especially our struggling Modern Western Civilization's. This is not just pure optimism, hope, dreams, wishes and will use sound logical factual structures, examples, evidences, proof, statistics, fragments, references, argumentations and will also use morals, ethics and emotions. This is an attempt to provide after each chapter primarily theoretical philosophical propositions and at the end of the book try to be unretortable on many Debate Issues.

This is to dispel pessimism, failure and despair which we have watched, listened to and experienced over the years, decades, centuries and millennia even though I am no more than a realist. This is to reveal multiple coinciding realities in this world of Power, Energy, Information Technology, Science and Technology ruled by money though there is at least a very strong presence of Open Source. This to regain constructive positivism which we have somewhere lost nullifying our productivity and profit leading Planet Earth into a downward spiral of deluge, sickness and disease, economical crisis and wars. This is to bring clarity, suggestions, innovations and solutions from Observation and Experience which is the distinct realm of Philosophy.

Nicht ontheil Zen Buddhism. Nicht ontheil Pythagoras. Nicht ontheil all of the Poet's, Philosopher's and Geniusses to date who try each and every day to improve the world.

A Philosopher's position is to reveal Truth. Truth is Commonality.

I, Kyle Lance Proudfoot, firsted started these essays at York University in Toronto, Canada (1995-1996 CE) where I got a 2nd year propedeuse in Philosophy and a 1st year propedeuse in Psychology in one year. After passing with an average of about 80% overall I moved with my Mother to Netherlands since her marriage broke up and I had no means nor money to stay and get a BA, MA and/or PHD though hindsight is always 24/20 and I would have probably at best become a Teacher and not have entered IT after getting my MCSE diploma and developing websites to date. So far I have become somewhat of a middle middle middle Celebrity, System's Engineer, Webdeveloper, Author, Artist, Actor, Musician, Poet and Philosopher. Life is a Sine Wave.

The last FREE Draft chapter at Silverlingo.com and Planesofexistence.eu was uploaded in 2013. See my curriculum vitae for other qualifications, projects and works.

This is Evolutionary Essays.

Intro Poem

The Animal

I am the animal,
The Ground and the Sky
And the Waters and the Sun,
I am all.
I feel the earth beneath my bare paw,
Swimming the currents,
Flying the waves,
The vision of the light I can hold,
I am the animal.
Mine is the current to make,
Eddy or whirlpool,
The matter is forged as I gesture
For all the Energy is mine to wield.
I am all,
I am animal.
The Elements,
Our builders of substance,
Essences of Form,
I am in touch with,
Like no other
I am.
All that matter;
I,
King of this Planet,
Ruler of all the matters,
Behold me!
I am Constructor
And Destroyer
With the might to determine it all.
Behold me!
I am the maker of my own Fate:
I am Human . . .

Aliens Blastin' Noobs

To conclude the predicament: None and All. For I have love in my heart and also believe in Theosophy, though as I define my system in The Free Show it accounts for no more than a fraction (like 1/32) of my entire Strength Of Faith structure module in it, which is the realization of the all existing God's and Goddesses and Philosophy's and Religion's, not just ONLY ONE single monotheistic GOD, for we each feel happiness in a different way and what is GOD without Henchmen who helped build our Civilization's, Planet's, Solar System's, Galaxy's and Universes. I state the last one in the plural too since these fall under different Planes Of Existence, Dimension's and Timelines within One Big Reality both internal and external in your sensory manifold from macrocosm to microcosm and from Finite to Near-Infinite to Infinity. I am not Anti Anti Anti your Monotheistic Philosophy's and/or Religion's which is so irritating everywhere these days from Noobies who think they know what they are going on about just cause they heard it in another Lie And Rumor Media Show, however as a fraction on Planet Earth in the somewhat blown 21st Century what would the actual percentage be . . . Also, when I have heard such about myself from them all I can do is hurl blow suck them cause ever since King Pendragon, King Arthur, Saint Patrick and King Edward there has been nothing but a Combined Celtic Christianity Complex for all your own CCC Complexes. This is, of course, just my own Right Of Choice (discussed in a later chapter). We see as a trend a moving away from such great extremities, unfortunately that brings Revolution and War with its upheavals through change. If you would bother to read even Islam and Judaism then you would also not be such an Anti Anti Anti Noob.

Before you know it Alien's will even show up (again) and what are you going to do, go and only Insult And Provoke them??! Well then, no one you get it right back in your faces. A Global Canadian Citizen allows for your Right Of Choice, unfortunately many systems being primarily still in Totalitarianism on both sides do not give you that and you have to Fight For Freedom just like we Fight For Peace all the time here in outdated crap living spaces with no insulations . . .

. . . and all their blast diameters, now you know the true meaning of Karma for there is no single Implosion and/or Explosion which goes in only 01 direction . . . Nicht Ontheil Einstein . . . if you cause that kind of disturbance then all the other particles on Planet Earth will go and show up and stomp that whole region . . .

To state here the other fractions of my own Philosophy's and Religion's is not the point and is actually NOYFB: Is this also not a Private Choice which is in your heart?

The following are some (hopefully) funny phrases too help loosen up your juices and put you at ease that I am not one of those highly over-complicated long drawn out boring old dry crusty Professor's at University's and other School's who put you to sleep.

'Name recognition is severely screwing my mind up. She's tryin' to lose her mind. Screw you too.'

'A Rose by any other Name remains the same. A Black Rose by any other Name remains the same pain.'

'They're tryin' to convert me . . . Is there Nothing in Everything? But about 99% of everything is nothing. Is it also 99% Light and only 1% Shadow?. Or is it a somewhat Spliffed Down The Middle mix. Are they just incoagulent?'

'Drunk. Free Will. Atheist. Science. Religion is like a drug. Then the Null came along: Zero Point Field.'

'Religion already went away in the Renaissance . . . Is it Israel: Checkmate. Maybe in hopes. PAY ME OFF. WAR Rules!! War is a necessity.'

'Thank GOD for Science!'

'We are the fighters for our collective warbird: A whole fleet of warbirds is certainly a Warbird.'

'A Dumb Shmuck wrote it down, way back then . . . , "The Aliens are comin'! The alien's are comin' . . . Panic now and avoid the rush!"'

'But then the rip came along: Alien's and Advanced Mutant's took control of the World!!'

'Near-Infinitum is close enough at unending .9's = Infinity = Immortal = We are all Immortal = To the length of the persistent databases of Humanity = Theosophy = I would sacrifice to him and/or her = LIFE and not DEATH = unending Life = unending Death = both = neither.'

'You are just a nanoscopic particle on a Near-Infinite Timeline.'

'What is wrong with this picture? Is it not the Ideas which have always shaped the course of the Timiline and is far greater and larger than your nanoscopic body?'

'It ain't blood I'm drinkin, that's for sure. Evil = Good. It's just a question of the beginning, the means, the method and/or the end.'

'Name recognition refers to the same GOD, who is defined as the Creator of all Reality.'

'"Kaduh Moron", psis the Alien. Holy Gruesome, Spy Kill!'

'There massive planetary guns open fire . . . '

'Humanity cannot save itself versus such a superior force with its present total lack of Interplanetary Defence System's, not to mention the total lack of Tank's on the continent . . . in fact, last we checked there's like only Russia and France with any standing Army left of any kind, America blown in failissement and China catching up rapidly . . . '

'The Civilization which had no Army to defend itself also had all its Art's ripped off.'

'Poetry Ad Infinitum, Defenc Ad Absurdum.'

'"Oh my God, oh my Goddess . . . they're comin' down in swarms, ahhhhhh . . . KABOOM!!!!" The last words of a brave Cyber Soldier . . .'

By the fact they are in quotes they are not per se my opinion. See Character Classes and Member's in The Free Show which has now become Spy Kill's 02 since 16-12-2013.

From A Cyborg Soldier To You

He was a young brave cop and decided to go there because he wanted to help. He was smart, talented, positive, strong, full of potential.

He was a friend of mine and you took out there whole Police Station with fully automatic weapons massacring 20+ Police Agent's in your terrorist psycho insanity.

By our right of vengeance and by your right of vengeance we will never show you mercy. We will unrelentlessly attack in legal and illegal warfare until we destroy you the demonic cockroach you really are. You are ruled by a lower pathetic Deity and are nothing but a Demon and I am ruled by Pluto and am a very powerful and energetic Mutant Telepath Telekinetic Cybor Soldier second only to Magneto. We spit on all your faces of your Demon and will rip them off sending you back to the blood soaked grounds of Hell Planes where you came from whether that be in a Higher of Lower Plane Of Hell . . .

I am a Demi-God Hero and do not tolerate your insulting presence. The very thought of your slaughterings by suicidal bombers, whole busses exploding, school children being ripped to shreds, your Terrorist Attack's against anyone opposed to your Totalitarianism, caught in the crossfire, your threats, your crimes, your pushing of hard drugs, your double and sleeper agents in our Territory's and Civilization's ready to strike at any time is an abomination to my mind: This is your Death Warrant for I and other Mutant's, and an already long existing unanimous agreement not to murder Christian's, Jew's, Muslim's, Women and Children and any other bogus Act's of War which are not in self-defense resulting in camps and death slaughterings of innocent Victim's, the cost of such per hour faillisementing our Civilization's, look at their thin broken gassed dead bodies, will annihilate your Evil off of Planet Earth and every last trace of a previous nightmare shadow of your existence; this part of Japan is not that part of Japan and that counts even double for all of your Country's in return . . . Nicht Ontheil Zen Buddhism.

DEATH TO ALL HOLY WAR'S. DEATH TO ALL TERRORIST'S.

DEATH TO THE BLOODTHIRSTY CANNIBAL FALLEN ONES who we will punish with the Power And Energy which we now wield: Null EM them with Laser Military!

LONG LIVE ALL THE ANGELS OF JUSTICE.

LONG LIVE ALL THE GREAT FALLEN HEROS.

LONG LIVE ALL THE TRUE AND RIGHTEOUS DEFENDER'S of Freedom, Peace, Justice and Free Democracy and the way of the Modern Western Civilization's.

LONG LIVE ALL GLOBAL CANADIAN CITIZEN'S. R.I.P.

The Cause of Changes in Evolution are Mutations

Solar Flares give us a clarification of the mutatious glory we experience throughout the course of Evolution.

Through Ra, the Sun, mostly random solar flares cause mutations in Life Form's on Planet Earth and now that Spy Sat Maya has scanned thousands of Galaxy's apparently also everywhere else in the Universe. I hope you don't mind some of my B—Stupid Violent Black Humor . . .

One can also find this proposal originally made by Darwin in high school text books.

By the variations seen in different Classes of Creatures it is plausible to think how wings grew through solar flares causing mutations in life on Planet Earth throughout the course of Evolution. At some point those wings of a reptile had to grow resulting in a bird and if you look at our own skeletal structure we are not so different, not that I want to try and flap flap flap off a hill or building again like everyone since Leonardo da Vinci . . . I find, after all, like many other statistics that Humor, regardless if it is not per se your Type Of Sense Of Humor (sry bad object description) tends to take the edge of and/or dryness.

In fact, is it not truly remarkable how almost all Animal's have brains, ears, eyes, noses, mouths, 4 limbs, a throat, stomach, bowels, muscles, tissues, cells, nervous systems and consume and make a lot of expletions . . . Could this be a mere coincidence and they for millennia long wanted to just toss our Pantheon's, Polytheism's and/or Poly-Animalistic Religion's and/or Philosophy's out with the somewhat rude, violent and uncivilized warfaring Celt's, well Druid's are none of such and rank there next to Pythagoras and any Roman Scholar who wants to take a crack at the whip. PS, out Celtic Civilization also never died out and they never conquered it past the Rhine in Europe and most certainly never past the English Channel into UK. Nicht Ontheil Bouddica and even Queen Elizabeth who did not fail in driving them back even though we are in 2014 definitely in the 2nd Renaissance.

Another example is to classify Group's of Animal's in circles and it is clear to see the photo of a handsome ape as most comparable to Human. It is also a fact that House Pet Owner's tend to start to (or already) look like their dog or cat. Since the beginning of Evolution with primarily horses we have been hand-in-hand with Animal's and by consuming Domestic and/or Wild Animal's we do not fail in ingesting their DNA.

Spontaneous Creation was done by GOD at Infinity, already, which is a Mut Point and really not a retort, and one cannot deny the development of Evolution from Nomad's and Tribal Villages to Hyper Modern City's with huge Sky Scraper's. One can also no longer deny the inevitable Space Race and colonization of other Planet's though now our estimates for even Laser Military, Mars or Luna Base are a half century because how long does it take to just get there and back with Mine Resources, allowing for the Hyper Acceleration Effect . . .

The argument of GOD is irrefutable. In other words, one can never deny the argument of the existence of a Creator. To do so, is to dissolve into an absurd over-redundant relational argumentation process of asking Stupid Question's excessively and continuously, like "Who created the Creator?" The Creator it/him/herself is already the FIRST ONE, the original cause and in their cases the original sin.

However, GOD ≠ SATAN otherwise we have no Free Will with the capability to choose between Good, Neutral and/or Evil.

Many variations can be seen in Life Form's on Planet Earth which are caused by mostly random mutations at different Time Point's and Timelines in Evolution for it is an interrelated Universe unto itself with many Power and Energy Pattern's characterized by the quantity and quality of Light and Shadow intermingling. Small, large and huge Sun Flares have the power to cause mutations, change, create and destroy life itself. Thank You RA! However, now we know, once again thanks to Maya Spy Sat there are also Black Holes in each Galaxy, thus these do not fail to also cause similar changes in Evolution by mutation and other disturbances and/or growth explosions which we see in 21st Century as being a Top 10 cause of the already half-barren devoured and/or polluted Planet which is very rapidly losing its self-sustainability for such a population.

In the development of Evolution are already many empirical measurements taken by Archaeologist's, Astro-Physicist's, Doctor's, Mathematician's, Philosopher's and many other Specialty's, especially these days Medias have their fingers in everything and everyone, as to the quality and quantity of Solar Flares, Environment's, Pollution and now the most horrific Natural Disaster's which is making all of the Prophecy's look bloody literal and not at all symbolic only. Solar Flares generate massive Energy outbursts which also have the Power to knock out whole power stations, networks, communication lines and many other things, not excluding your very own brain and nervous system which like at Full Moon reacts to EM Field's. Could it be as some have said in Medias that the actual cause of all of these things is no more no less than the feared and dreaded simultaneous over-population explosion of multiple Races and Species . . . This position is actually supported by the vast majority of the communities.

These heightened Power and Energy Level's are measured even by Space Agency's, a 'Richter' scale is a good analogy but pales in comparison to the absolutely incredible quantity and quality of Power and Energy being generated and released in the 21st Century. Never before have we had Internet and so many Technological Devices.

In 2003 there were also larger Solar Flares. Cyclically, stronger Solar Flares are a regular phenomena on Earth. My personal opinion, since everyone due to my own Parent's and Vegetarian past from 1988 to 2001 seems to want to try and label me as a greeny, I even donate to Greenpeace, is that Global Warming is an oxymoron with only the exception of the Arctic as proven by Al Gore, the rest looks actually a lot more like massive imbalances and disturbances of the entire planetary ecosystem; I'm actually not green anymore but all the shades of gray in the Universe, I'm also not even a Half-Vegetarian anymore but do eat free range and/or organic meat so they at least have a bit of a life and are not pumped full of hormones and/or antibiotics and/or all the other spinoffs of how fast can I make an irradiated genetically modified clone sheep grow . . .

DNA also interacts regularily with its environment through the Power and Energy Field Level's which perpetually exist and fluctuate; as seen by the recent Ice Storm's in Jan 2014 in NA we could also suffer another borderline Ice Age of some kind. DNA can be seen as a kind of blue-print, road-map or Information Highway of the Creatures on Planet Earth all interacting, eating, killing, mating and procreating each other. Even our DNA can be bombarded by different wavelengths of Energy, Light and Shadow Energy being, probably, the most powerful Form's of Energy! After all, once again, look at the relative size of our Solar Sun to Planet Earth itself and that of the Black Hole . . . The Speed Of Light is presently the fastest speed matter can travel, however Shadow Speed will soon eclipse it once they have various breakthroughs as depicted in Science Fiction/Fantasy Film's and 3D Games. Many now also support the Theory that Space Time is actually a fluidic continuum of EM Field's. I find it a little hard to swallow that there are no particles because there is still the Wave Particle Theory. In any case, next to the obvious Chaos Theory taking over and dominating at this time, hopefully somewhere like a Game of Solitaire the Law of Schrödinger should not fail . . .

You got to ask yourself though: How long does it take Karma to kick in now and zen . . .

Based on the facts to round off: GOD created Infinity, there is no denial of the development in Evolution, Solar Flares cause mutations in Evolution through Life Form's, stronger ones are a regular phenomena in the Universe, there are related variations to be seen in all Life Form's on Planet Earth by simple DNA comparisons, it can therefore be concluded the cause of changes in Evolution are mutations triggered mostly by solar flares and at this time still highly theoretical Black Hole influences.

Potentius in immortalus ad n infinitum sacrus solarus ex shadowus.

(In case your wondering, again, like so many to date, why the capitals, well next to the fact English comes from German, those are also my Object Name Convention's as defined in The Free Show and other parts of my works, that's what you get for not finreadin' me . . . = Noooooooobie!)

Politic's Is Disturbed Too

Politic's is disturbed too.

Yes, but at least Politic's has a practical function. There are so many people, especially Politician's, who only know how to bullshit and do too many Personal Attack's; after you've lived in The Hague for as long as I have and at least watch some New's and/or Politic's on TV and/or Internet instead of what the blow statistics state that there are only 2,2 million people in entire NL who watch the New's with 100+ Reality Show's and other forms of Entertainment dominating the Medias then you become a bit of a mini-professor in the Topic.

One of the greatest errors in Politic's in History Of Humanity to date is to screw up your premises, arguments, facts, references and/or statistics: There is nothing better in Debates then to come up with a really great retort even days, weeks, months, years, decades, centuries and even millennia later, especially right before Election's is devestating to the Opposition because they do not have enough time to counter-retort. Also very erroneous is to not at all argue Issues, especially in the 21st Century which is driven by 1000's of Issues since people, at least in Modern Western Civilization's have the Right Of Free Speech instead of being thrown in dark dungeons and other horrific tortures and prosecutions throughout the millennia by all cultures. You can then just as easily be targeted right back for doing such and having the worst unfounded bias against i.e. Bill Clinton again, since I swear it to I never had sex with a 13-year old but as a kiddy myself I did not fail to fool around with my own Friend's . . . oopsy, now they use that to against Britney Spear's and walk away with the Election, Grammy and all the $$.

I also consider a 1, 2 and only 3 Party system still stuck in the outdate Left, Middle, Right when Issues are far more complex, and these days on Internet if you do not deliver my product in 24-48 hours I just step over to the competition, to be far too simple leading to bottleneck wrench blockage Spliffed Down The Middle Effect's. Our system here in NL is however then far too diverse leading to fracture fraction fractal fucture displays for 12-24 Party's leads to nothing but bad voting i.e. 2012 Election's PvDA could have even won if 15 seats did not still go to SP which would have entirely changed the whole NL, there was also only a difference again of about 2 seats between VvD just like in late 2013 when NL finally reached an accord with exactly 1 seat difference . . . Well, nothing will change and/or improve therefore as the go and block everything just like with Obama and Republican's until they to pulled off a debt ceiling House uproar upheaval of the Republican's. And then unfortunately, his Obama Hellth Care system hit even more wrenches and bottlenecks from paragraphs stuck in the Colonial Ages just like their suffering Education system which now has no chance because there is World Crisis.

So, where is the not unhappy not imbalance . . .

Social Democracy and/or Absolute Democracy and/or Hyper Modern Western Democracy are in principal the best choices, the least of all evils but they do not exist! According to the MySQL database of 2008 there is NO Democracy, yet. America is a Federation and not a Confederation so they have the full right to blow each other the Hell away. Canada is a Constitutional Monarchy just like Netherlands. France is, obviously, a Republic though with Hollande it's trying to call itself a Social Democracy because it is somewhat Left. In UK a middle ground was found with a sense that each side got something out of it, except Scotland and Ireland who want Independance. In Europe the sense of cultural pride and honor and Identity and self-autonomy has never been stronger, what is this again: Social Nationalism. I am actually getting somewhat confused now considering WWII. What are we then?

These political systems are best demonstrated by Netherlands, Canada, UK, France and America. They allow in different ways for the best stability and flexibility despite their drawbacks. The worst example of it are now the so-called unification of systems into blow groups called EU, EC, EER, US, USA, UE and all other such failing attempts; people actually all just still hate each other's guts and conduct the worst type of competition, wars and hacking.

'It's like they don't want any more International Trade.'

'Talking about missing the Tourist Potential completely again . . . '

'Let's all shoot, kill, rip, rape, murder, pillage first and never talk cause now I own you.'

'If you don't even have a Knight's Round Table to talk with each other then you also have no First Instances to talk about.'

Social Democracy certainly does not lead to vast accumulation of materialistic wealth by some individuals, however it does protect the individual a lot better and it does lead to a spread horizontal distribution of Wealth which is necessary for those less capable. Unfortunately, the taxes are presently so bad up to ala 75% of each and every euro going to the Axe Man but in 2013 statistics stated that +/—156,000 people leave Netherlands each year, like some kind of mass exodus fleeing for some reason or other which is not per se the Continent being run over by Immigrant's again, which is an Anti-Position, but the taxes alone are enough for you to just pack all your baggage and move to France. Aren't we technically paying more taxes than Russia who are now coming across as more a Left Labor Party even, with the exception of some incidents, and not at all Communism. The Chinese as my Uncle David likes to joke is actually Capitalistic Communism.

In some ways, Social Democracy is the anti-pole of Capitalism which is why both camps, especially in UK and America hate each other's guts. In other ways, it agrees, for without Economic Evolution there is starvation; proof of pudding: Go up to a Northern European these days, for North Europe is not even the same Planet as South Europe anymore, and while staring at their bulge next to their V-Line ask us if we want to donate, invest, be ripped and/or raped by your far out in left field project again . . . Like I said, I'm not Anti anything except the worst extremes but what do we get out of our participation in EU from the Greeks after having to donate Bernard's to them after they were fucked in their Bank's by you know who . . . Do we own anything there? Do they love us? Are we best of pals? What a bunch of malarkey, at next Election's even Merkel in Germany will probably be tossed into some footnote, though their SDP did not fail to win and she remained at the top . . .

Absolute Democracy is the strongest type of Democracy but not per se the best choice which allows for fairness to all Citizen's at all levels and is realisable with voting on Internet, but unfortunately: 'Hacker's wreck it for each and everyone all the time.' If the people could actually Vote on many Issues at every level through Internet, even daily, instead of only once per 2 or 4 years with the Classic Representative system then the true desires of the people would be known and possibly realized, on the other hand it could also lead to the worst kind of Chaos and Anarchy since Internet is now notorious for suffering the most from such. However, given time, it might be doable in some smaller more isolated Island Mentality's like only your own business. For example, in the future when there are Dome City's I would want to vote on the weather. Also, another major trouble with the Representative system dating back to Roman Senate, Greek narrators who got kicked off stage by their 'dictators', Locke and the American Independance is he and/or she is 'alone, weak and vulnerable' and can be easily attacked, bribed and/or blackmailed if not killed, murdered and/or assassinated which we have seen is a frightening growing trend since the Martin Luther King and the Kennedy's though it is, obviously, not unheard of in entire History Of Humanity, and these individuals are often far too susceptible to the Power's and Energy's which be, especially the still to date opposed tribes of Warrior and Priest. However, don't we need Medic's in our Army to regen and heal the wounded if not even reanimate . . . I could also argue in the 21st Century with Information Technology deciding each sector that the better choice is Wizard and Warrior, otherwise why not just let them all invade and rip and rape not just your Art's and Sciences but all your Women and Children too . . .

Hyper Modern Western Democracy is probably a good middle ground between Social Democracy, Absolute Democracy and Capitalistic Democracy (which hardly needs a description after 5 Wall Street crashes in about a decade and the rise of the East) and is far superior by definition to the total lack of Democracy in the eastern blocks and pretty much the entire southern hemisphere of Planet Earth who come a lot closer to outdate Despotism's, Monarchy's, Dictatorship's and Military leadership's, all once again another mask of Totalitarianism.

Hyper Modern Western Democracy would entail taking the best quantities and qualities of each of the present (and past ones for examples and material) Modern

Western Democracy's so we can solve serious problems and imbalances which are sending our Civilization's into not only faillisement but spiralling down and out of control as we see in many Medias millions of people keeling over dead from one thing or another left, right and center, thus why not also the Dead Centre Shopping Mall . . .

However, the drawback of putting too much water in the wine is turning everything into vinegar and risking too much middle ground and compensation by making almost everyone just unhappy i.e. if we see anymore basic self-contradictions and/or false promises from Party Leader's and then all of a sudden it's even the total opposite in no time flat you not only lose the trust of your people but what *does* your Party stand for?

At present most of the world population has serious concerns about its House, Energy and Food supply with the worst Inflation Rates and Rip Off Euro Zone in North Europe since, yes, again, WWI-WWII. This is Evil at its best which then stauntly defends its position in favor of competition and wipes out all those opposed. Also, excessive population does not fail to also lead to hyper aggressive expansion policies.

Cheating, backstabbing, hacking, attacking and conducting Battle and War under the guise of competition is based on a false super ego complex and you have little regard for the value of 1 Human's life, it remains nothing more than the age old Take The Money And Run, I now have your Contract and with this much bloodshed of African's alone I now have all your Territory and Resources. There I win, shut up, I own you so you it's not like you can argue and/or Debate the Issue; we have that quite badly here in NL with the classical way to many Rules, Regulation's and Law's where each and every square centimeter, if not nanometer with Hyper Modern Civil Engineering (smirk, don't take it personally or anything that's just my sense of B—Stupid Violent Black Wireless Humor, because how can we decide such ownership), is accounted for.

In Capitalism, if you ask me there is nothing wrong with Absolute Capitalism but it got so badly watered down and also sold down the river that you can just shred the whole contract on this side and they are actually stuck in a prism of their own design having signed 50-100 year contracts to the Arab's, 1 CEO, 1 President, 1 Vice-President, 1 Head Of Operation's or 1 Manager can decide the fate of 1000's, if not millions, of Family's: History Of Humanity is wrought with these Labor disputes. This is a greater 'number' oriented system with emphasis on profit and not personnel. There's very little concern for the life of the individual, strangely enough, considering they're the ones who have to do all the work for you, or a handful of Stock Holder's, or just it itself the Dark Evil Sky Scraper Complex from West to East Coast when in the future there is only one left over after all the cut throat competition has destroyed everything in between.

There is nothing wrong with per se a Hierarchal system, this is unavoidable and is present throughout entire History Of Humanity, after all someone has to own and if you give it to the people to Vote on another EU Referendum they'll hurl blow suck the whole system with a big NO and probably burn the whole thing down by accident again. The Future of Humanity based on Human's innate nature and all of its good intentions, we're not per se Good, Neutral and/or Evil we're just exceedingly uneducated, inexperienced and primarily ruled by our Need's, Want's, Desires and/or Emotion's with the exception of what percentage of Human Species again? Stick 2 decimal places on even 100 million only and you're so way off that you would destroy entire Planet Earth.

A double checking self-redundant system could be better to counteract the great imbalances of Power, Energy and decision making done by present Politician's and Economist's, Business Owner's and Leader's who practically have Absolute Power and Absolute Energy over all things: Present graphs show that +/—1.0% of the entire Human Species are the Rich Elite. This has technically never changed in History Of Humanity so is not only a 21st Century phenomena and it will probably never change . . .

Hyper Modern Western Democracy would probably lead to an unending amount of disagreement as everyone fights for a piece of the pie never leading to even some kind of Universal Unilateral Global Democracy. Oh wait, it's now NWO in 2014, oops already.

Absolute Democracy would probably crunch to a halt with a ridiculous quantity and quality of daily voting with your cereal, not to mention a staggering increase in Internet hacking and it would dissolve into even worse Chaos and Anarchy, read Eric Schmidt.

In Social Democracy, you as an individual, are protected by Social Insurances. With higher taxes you cannot acquire the same level of individual wealth or even in my own experience in NL being now on a lifetime disability welfare check never escape from it either, ha ha ha, exactly 30 years later 1984 finally comes true . . . but if you get sick like OHIP in Canada you don't have to blow all your savings and end up on the street amongst millions of other homeless or stuck up to your neck in 6+ Credit Card's.

My bias is, thus, in favor of Social Democracy mostly because there is less imbalance; the differences between rich and poor has caused the most Evil in History Of Humanity. In fact, a highly materialistic wealth acquiring profit hungry Predator always feeds on the lesser idealistic wealth shunning profit sharing Prey. It's even in our genes.

Therefore, what is your own Vote, your own Work, your own life in Democracy today?

Later in Evolutionary Essay's I propose a whole new system called Free Democracy which attempts to take the best elements from each system on Planet Earth.

Null Religion

This is the reason for Null Religion.

Your Religion's are primarily based in bias and superstition and are caught up in abstract symbolic vague incromprehensible gesturing through rities and rituals. Your symbologies alone takes about 2 centuries to decipher per book. No one really has any clue what most of your paragraphs mean and there are this many interpretations of each of your paragraphs of each of your Religion's, and this is one of your daily errors with all of us in Science, Technology, Atheism and Information Technology for we are not talking about ONLY your Religion, since you react to this so vehemently each time you must be guilty as red . . . This is the reason you have all your Holy War's, Death, Strife, Prejudice, Sacrifice and Suffering. This confusion, abstraction, over-complication, vagueness, misinterpretation, mistranslation, old school to young school, verbal tradition to written tradition 2 centuries later, originator and inheritor and all kinds of other errors from Scholar's to Peasant's through watching, reading and/or listening from their own subjective glasses has actually caused what is the opposite of what your Religion's actually stand for, thus the out of control split factions and all their conflicts which kill everyone and everything in between. Your problem is you shoot first and talk later, they then shoot back, I am also only Switzerland myself too with a Fire Back Only Policy across my i.e. computer systems but now to date as some argue everyone over the millennia has already fired and shot back all the time so know one knows what is Original Sin.

GOD gave you also a brain, not only bwain damage which is how it seems most of the time due to all the suffering to date. By that fact, GOD also gave you Rationality, Reason, Intellect and an IQ Level through your bloodline and parents. Therefore, the strange misnomer these days spread by god awful Ignorance that your Religion's are opposed to Science, Technology and Information Technology is the worst Lie Of Satan to date.

However, it does seem quite evident that Atheism and Anarchist's are indeed opposed to your Religion's having looked at the millennia and decided to simply be with ONLY Science, Technology and Information Technology which they do not lack the Right Of Choice in and should not be perceived as Lucifer worshipping heretics like so many have been perceived by your Ignorant Superstitious Uneducated Peasant's to date, unless of course they really want that image.

Still though, a good Lesson we have learned from the previous regime of the Vatican and their so-called post-Mussolini era is that your best choice since you are definitely opposed in Alignment is to just not Interact with each other. This is now also proven with statistics from Israel and other Country's.

Otherwise, it's no surprize that you keep killing each other; all it takes these days is to jump up and down go 'ooga booga' and call someone a neo-nazi or gay voyeur and they charge madly attacking; it's too late if you are now dead.

Most Religion's, not just Monotheism's which is putting a bit of water in the wine from me as a Celtic Baptist Christian, are bringers of false hope and do not actually in reality put bread, meat, vegetables and drink on the tables of the population. Claims of such come across to us as more like casting Magic Spell's and if he would actually raise his arms once to the skies and call upon the real Divine Magic without using Money and/or paying for the table in one way or another but that it actually appears just like Jezus Christ was able to perform all kinds of Paranormal Capability's then we would follow him even in a 1/8 at a min fraction of entire Planet Earth. This is also strange to us that we in Science Fiction/Fantasy and other genres get so much criticism and superstition, like oh no we're ALL Evil again, when your very own Prophet could do like every Paranormal Capability. Regardless if it comes from Divine Magic which is still also our Celtic God's and Goddesses and GOD always above always looking down or from Arcane Magic's, the simple difference between Priest and Mage make no difference to the fact that it is some Type of Paranormal Capability which does not use any Technological Devices. That to me seems more the core of the Issue but that is high abstract theoretical semantical bullshit even and has nothing to do with the very not unreal conflicts and oppositions which have theologically, theoretically and technically actually no basis in fact or truth.

Honestly, if we look at History Of Humanity in a new Light and Shadow then we see who really did them all in as the 3rd Party Antagonist: Who sacked Rome again? The Goths. What is the direct bipolar opposite of Sun, Light, Truth, Love, Wisdom? Black Hole, Shadow, Lie, Hate, Ignorance. This to us exonerates the Jew's completely and their Enemy's to date are nothing but filled with Hate for them and some are even possessed by Demon's having turned into Blood Cannibal's, again, who would have though WWII would repeat itself, and just slaughtering everyone.

What is now interesting and that is why this chapter has this title is that we are wondering if all of such developments in Timelines and Alignment's in systems line up directly with Quantum Physic's of Black Holes and Sun's interacting with each other and the Planet's, Moon's, Constellation's, EM Field's and Ideologies on Planet Earth.

Null stands for Null Point and Zero Point Field the mythical center point of each and all EM Field's which could potentially lead to Infinite Power And Energy and also like the Atomic Bomb lead to an invincible weapon. Or, as I like to state, at first they will use to create weapons of mass destruction but then it will also provide Near-Infinite Defenses.

My Title And Status is not invented, I got a Baptist Priest to dope me when I was 14-years old in Toronto, Canada next to Noah's Health Food Store and I do not lack my bloodline from the Clan Mac Aulay which may be connected to McCollaughs (15ᵗʰ Century) and McColloughs and McCollons plus other Clan's and spelling variations who were Royalty—Dulce Periculum—Danger is sweet—Gaelic name: Mac Amlaidh who are probably descended from the Ancient Royal House of Monster, a Norse Clan, and a Gaelic Clan. In about 1200, a son of the Earl of Lennox was probably a direct ancestor—Crest Badge: Antique Knight's Boot, couped at the ankle which is practically identical to the Proudfoot Crest Badge, the last name I carry today and who emigrated to Louisiana in the New World in 1777 CE—I also have the middle name Lans which was translated at my birth in Toronto, Canada to Lance and who in the 19ᵗʰ Century had an Admiral who was granted Royalty. It probably goes further back even . . . I simply did not register myself since back then it was not mandatory to register your Real Identity into a Church. I did this to stay below the radar since I was a strict Vegetarian back then and I did not want the attention. That didn't last for long and when we moved to Netherlands (see previous chapters) I got lit up like an FM Christmas Tree. So, do I get Titles and Dowry?

This is exactly a part of my point, due to prevailing Ignorance and also by the fact I had a daily marijuana habit and a bad temper which I still have to date I already got misintrepretated from Step 1 of the bat, kerswing, wham, bam, thank you mam, goodbye behind the Parni Banger Slammer. However, don't forget for all of your own Insult's by almost all of you directed at me that I am nothing more than any of your own Celebrity's with a couple Habit's And Tradition's, i.e. read first sign of State Of Denial, Ostrich Effect's and/or Dump Sand Over It All equals addiction.

Religion like Science and anything else like chocolate can also be an addiction. But to point your finger at me or us and say that Beer, Wine, Alchohol, Marijuana is worse than White Salt, White Sugar, Chlorine, Cocaine not to mention, 'Stop that I know you're eating a piece of chocolate or chips now!' with all the highly over-moralistic Do's And Don't's like your smartphone giving you warning blips every 15 minutes is not really the point of your and/or my Religion. Religion is meant to develop the religious, spiritual, philosophical, compassionate, sympathetic, caring, loving values of Enlightenment, Rationality, Light, Truth, Love, Wisdom and even Knowledge. Now, here we have the basic self-contradiction misintrepreted by so many. Since when therefore is Science, Technology, Information Technology and even all others who seek Knowledge, Intelligence and Wisdom (the latter I never developed myself) opposed to Religion?

Quote Socrates: 'The only sin is ignorance.'

Don't you mean you are opposed to your Enemy, those who would dare break down your City Wall's and Gates and rip and rape all your Women and Children, steal all your Gold and Silver and then camp out amongst all the dead bodies, decide it's too much of a bother to clean up all the wrecked decapitated broken Roman and Greek statues, leave and then after another War give it all back to the original proprietors . . .

We don't comprehend this phenomena in History Of Humanity, what is this urge to kill, murder and assassinate each all the time but some form Satanism. Lucifer is probably the Son of Satan and not Satan himself since there Element's do not line up: The Son is NOT like the Father in many instances. So, they took their revenge from before written History Of Humanity in the Great Old War with the fall of Lucifer and sacrificed Jezus Christ. This is to be found in an obscure text from the vaults of the Vatican which few have read though you might be able to find it somewhere . . .

However, our Religion's and Philosophy's are not the Scientific Method.

'Thank GOD for Science.'

Nor can Christ cure Cancer. Nulling on the other hand has much potential but next to great Comic Strip's and Hollywood Special Effect's and Stunt's is there anyone on entire Planet Earth who has a single real genuine Paranormal Capability i.e has anyone levitated off the ground up to even 200 meters in the air with no help at Times Square? No, most certainly not otherwise it would have been all over International New's Station's that the Mutant War's have started . . . Cute little stunts at a couple meters only can be pulled off even easier than Praag Radio Ear-Plug Devices.

Look to Science and Technology as an alternative to Blind Faith which to me is far worse than taking Blind Order's, doing Window's Blind's for not cheap thrills, being Blinded By The Light and having the One Eyed Lead The Blind, again. You can spend 5000 lifetimes on Science and Technology alone.

Religion is not based in Logic and Reason though it does teach Rationality, it is based in your Strength Of Faith which I define in fractions, so in other words by your Right Of Choice Of Religion's you don't have to choose ONLY ONE. Unfortunately, other Country's do not grant you this right so it is still per Country and not per Planet.

Strength Of Faith and your belief and your motivation and your loyalty and your dedication and your commitment is very hard these days for we too see only death and deluge with the difference between rich and poor becoming even bigger. I, myself, and me have never been to good at this one which is maybe where a lot of the confusion and conflict with other people in my life came from unless it is staunchly defending my own, however I tend to unfortunately do that by cursing you all to the Hell Fires of Lucifer and the Planes Of Hell; I like to state these days: 'Pick a number between 1 and 666 cause that's where I'll send you to, never meet you there . . . ' or 'Prolongation of suffering? Why not give 'em a good kick in their asses and hasten their demises off to the Abysses . . . '

'I do not condone and/or condemn Lucifer, I dump him on your heads . . . ' This is a problem word in prunounciations, if you mispronounce it by accident in a sentence by accident or it gets 'ripped up' then they freak out and another National Insult has been committed . . .

Science is a Religion. Well, the first steps in Modern Western Civilization with the Pre-Socratic's, my Specialty in my 2nd Year Philosophy at York University, Toronto, Canada, put all Specialty's into one big Topic called Philosophy. Only until much later did they diverge off into disjointed departments called Specialty's who no longer even talk with each other, another classic example of how in our own Government's and Corporation's that the Left Hand has no clue what the Middle Brain what the Right Hand is doing . . . This seems to me, the hammer on the head of the hammer whale, not just the sperm whale . . .

If, once again, you don't even have a Knight's Round Table to talk about such Issues then you also have no First Instances . . .

If your Organization, or thus these days the total lack of such Order, does not even have a PR Seat then what the Hell do you expect . . .

'Something must be decided with logical argumentation, morality, and plenty of Romulan Tactic's.'

The vast majority of decisions and problems are made by emotional choices and not at all Logic, Reason and/or Rationality.

Well, at least we have this in common since Science, Technology, Information Technology, Religion and Philosophy primarily ruled by the left hemisphere of your brain actually abhors all of those chaotic firy and watery emotions of the other hemisphere. Cute how that lines up with the planetary earth geographical systems . . . Could we be on Mother Earth' head or is she in circa 2013 at End Of Cycles suffering the worst kind of menopauze and through such Natural Disaster's offing the parasitical infections . . .

I'm not Anti-Woman either, and since a lot Priest's are Homosexual and/or Bisexual how could they possibly be Anti-Woman in the First Instances? I just can't resist that Black Humor, not just because of over-population, but because a previous Gothic Girlfriend gave me a yeast infection for 6-9 months which I also gave to the next Celtic Girlfriend unwittingly and now I have scar tissue for the rest of my life . . . That is why, next to blow hurl suck Rules, Law's and Regulation's about Marriage and Inheritance with Rip Off Euro Rates that I don't look for anyone . . .

All of this rambling has a point: Is it not just like the Education System stuck in the Colonial Ages the fact that each of our Religion's and Philosophy's are stuck in century and millennia old scriptures, texts, fragments, tenets and laws? And if we try and update them, we suffer the worst Violation's and even sacrileges. This is the same X-Machine computational matrix mathematical problem of: Who gets to define the point system?

The worst example to date of such for me was from my late Ir. Oom Han (R.I.P.) who stated on more that one occasion that he is totally for an Enlightened Dictatorship and a Weighted Voting System. Well, he was great in Science and Technology, but I'm sorry with that State alone in America if you gave Obama 1 million votes it would win the whole Election's and if you gave Putin 1 million votes than he would stay 8 Stalin terms.

It is just is what it is for it is developing through the course of Evolution and that is also what I think of all the blow hurl suck IT we've to date seen; once again, the most common cause of changes in Evolution are Solar Flares inducing random mutations and IT Network's are especially vulnerable. Intelligence and IQ Level's are the primary factors in the survival of a Race or Species which is why Man and Woman are the strongest, fastest and smartest Species. This is primarily Logic and Reason to the extent of a factor of 99.9% to .1% and has very little to with belief; once again, everybody and everyone has their own sunglasses on, in their cases its blatant myopic tunnel vision.

Also considering present day circumstances and how fanatical Human's apply certain Religion's and Philosophy's, it is clear Logic, Reason and/or Rationality is a better discourse, especially in 21st Century as we are in a descent straight into the Hell's.

Give me Null! Give me Null EM! Nulling is the metaphysical middle to Paranormal Activity's and Quantum Science exploration. It can theoretically even lead to Infinite Power And Energy which then makes it interesting for the 1.0 % Rich Elite. Regardless of how I like to play with the people or not as an Anal Intellectual one cannot deny the top to bottom hierarchal structures, those who do so are like we use to be as overly optimistic and idealistic Children and Teenager's with all the false hopes for the futures.

If there is a desire for some form of Mysticism, which is included under Philosophy though some say it should be separate, then just as with your present lack of rights you cannot fail to make that choice in your Soul, Spirit, Mind, Heart, Body, Penis Principles.

Mythology is a great source of inspiration and I am still gripy that 'Conqueror rewrites History' that is somehow not an official Religion, it shall always be a Philosophy.

However, Realism seems to be the Rule Of The Day and Blind Faith is dominant. Blind Order's through the motto of The Free Show 'Obey or Die!' which is Black Humor also dominates only since you can also not say 'No!' to your Boss or 'Get Fired!' Kill Or Be Killed and Don't Get Caught In The Act are the only Law. Go and murder in the name of your Holy War if you believe that is Justice but regardless of what your definitions, interpretations and/or beliefs are of the After Life, it still exists . . .

So, I, a Mutant, in the Mutant Generation, choose for Null Religion, for lack of proof . . .

The Necessity of Morality

To make a decision you can use Logic, Reason, Rationality, Emotion and/or Morality.

In shorts a Fool, always write first your name in your book and that which is uppermost in your Heart (Edgar Allen Poe):

1. Logic is 'number' oriented like you are only a Planck Number.
2. Reason is based on 'argumentation' like if you drop the cost of a software package by 12% then you also sell more with the Bell Curve Theory.
3. Rationality is based on a 'system of thought' like in Religion's, Philosophy's and Psychology's like Aesthetic's, Asceticism's, Altruism, Reductionism, Group and Individual Theory.
4. Emotion based choices are 'intuitive' like what you perceive from yourSenses and Feeling's in your own subjective realities.
5. Morality is based on 'ethical' choices like Environmental and/or Animal and/or Human Friendly.

Very important to the necessity of Morality which according to ours and many others' Death Analysis Trend's on Planet Earth in 21st Century which is mostly blatantly ignored and written off and scoffed at as 'Far out in left field green freakdom positions' due to their unending Republican Anti-Left Anti-Green Anti-Anything-Which-Is-Not-Us who actually provide no jobs at all and/or self-sustainability as is proven by the minussing Economy's which they had no small part in and at this rate of over-consumption by 1/3 of the population of Planet Earth of 2/3 of its Resources with their ever-stubborn refusal to get the Hell out of the Fossil Fuel Age, we'll probably all be pulled down by them, is the Fundament's, Foundation's and Methodology's of applying such decision making from paper to practice. And as they say, I'm not only blaming the Republican's, but in the famous words of Ben Tiggelaar if your business has 12 Department's and your order has to go through them all to be processed and/or get permission then no wonder you're already losing money heads over heels . . .

If you look at each of your own choices then you see how it went horribly wrong:

Examples of how badly we each screwed up and why the East is taking over:

1. Deciding how to most efficiently run a company or do you have a paper chase.
2. Primarily people and humanitarian oriented and no plant or animal orientation.
3. You the CEO or Manager only decide what is best for the situation, the unique relative circumstance with practically no input and/or voting from your colleagues and/or employees and/or consumers.
4. Did you use primarily Logic, Reason, Rationality, Emotion or Morality?

And equally so for other combinations. So many times in Debates and/or Battles, pick in each Session a fraction, recommended is 9/10 Debate and 1/10 Battle, I have encountered Noobies and No Noobies who do not have their premises correct. If you do not do your homework, enough Research and Development and/or testing then I am going to show up and blow your version of your software package the Hell away i.e. I'm sorry ok, like I said on multiple occasions now I'm really not *that* Anti-Apple as so many camp so hald these days, but when the acronym SRI does also not fail to line up with RSI, IRS, SIR and RIS like my previous joke on it and everytime you say 'fuck' it says, 'Stop saying fuck, John, swearing is not good for your Hell . . .' then all I'm gonna do is get more blood pressure and hurl it against a wall across that many wave files uploaded and downloaded back and forth . . .

'IF logic was used to state it is necessary to use both THEN is it actually a logical decision process making need and/or want?'

This is a very abstract way, if not somewhat bipolar dyslexic like most of our decision making processes, of stating it is impossible to use Logic, Reason, Rationality, Emotion without Morality and vice versa. For, you cannot in your decision making ignore the needs, wants, desires and emotions of the parties involved. If you simply say, 'These are the numbers and there are no other options and that cannot be done' and leave it at such alone like we have experienced countless times to date by all sides of the spectrum then you have reduced everyone to a mere 2-Bit Useless bit of info and you might as well only go Nicht Onheil Heisenberg and Heidegger unending while acting really smart like one of those highly over-complicated Hyper Nerd's who cannot make a 3 layered file folder system to save their lives and systems and then oopsy you still gave everyone only a 6 digit postal system number and caused another Millennia Bug. Meanwhile everyone has gone faillisement due to Waste Time Effect's like all those god awful printers to date and everyone has keeled over dead left, right and center by default from swing of the bat cause what the Hell is now a Factory Setting anymore?

Morality being an integral part of Mythology it is a good substitute for Blind Stupidity.

'Never underestimate the stupidity of their Humanity.'

'Oh no, now they killed all Art's, Sciences and Humanity's again, not just rip them off.'

Morality means you behave according to a specific code, not do Improv Politic's, Improv Economic's, Improv Dildos and now they also want to do Improv Warfare!! Whatever happened to all the rules, etiquette and laws of Battle? That isn't even ordered regiments but blatant mob anarchistic riots. Maybe you're better off doing a Fighting Withdrawal regrouping, doing some standard training and wait for Reinforcement's, do you really expect to win if you do not have a 10000 Army . . . we don't think that those paragraphs mean just blatantly offing yourselves by self-suicide, should you not try and take at least 1000 with you . . . here in northern Europe we have a similar code: If you think you're going to end up in Valhalla with your own privileged grave position and you go up to him and/or her and say, 'I threw a stone and got shot through the chest by a tank, do I get to make it to Valhalla?' then he and/or she will reply, 'Why don't you go through that bright door there and bend over and sow the seeds in the fields . . .'

Moral's and Ethic's works also in combination a Behavior Code though these days it's coming across more like Behavior Police, Thought Police and Mind Control, I would rather have the remote control to that Mind Control Machine . . . Instead of trying to Bwainwash them all the time, why don't you try to Bwainstorm them . . .

Morality and Ethic's means the upholding of happy, positive and polite ways of communicating though no one can hold that anymore for more than a minute either, myself most certainly not excluded cause I don't exclude myself from the equation unless the whole stage explodes on fire, I jump off dead center with my butsky on fire and I run out sprinting and screaming through my own twirly dark black shadow double whirling doorways, the whole theatre explodes and everyone dies, what a great Romantic Tragedy, Holy Gruesome, Spy Kill's . . .

So, if someone you don't even know walks in fanatically blabbing away about something without a single discourse to Logic and/or Reason and/or Rationality and/or Emotion and/or Morality than you have the right to give him a right middle finger. It, however, does not allow you to say such is not your God so blow everyone and everything the Hell away, again, and this many times to date and to come in History Of Humanity. Justifying killing, murdering and/or assassinating Group's and Individual's by trying to argue any of these things except in Self-Defense is in violation of almost each and every system in Existence except the worst militaristic dominating types. And as we see, since they are now in the Weak Minority on Planet Earth all other systems are banning them.

This as with most systems does not work in all situations where recourse to Adaptive AI is necessary i.e. we have still not accounted for a vastly superior Alien Species with no Laser Military Planet Defenses who just don't give a shit about any of our systems and wipe the whole thing out just to clear it for the Resources and/or Interstellar Highway.

The correct balance of Morality with others is demonstrated with the interaction of Subject's, Object's and Values. The right hand must always be balanced with the left hand and the middle brain . . .

The Purpose Of Evolution

The purpose of Evolution is to develop more Energy and more complicated Energy interactions. Energy and Matter are interchangeable, therefore one can lead to the other.

If you look at a dog or bird compared to a Human then you can see they occupy themselves with far less complicated tasks, they also sleep a lot and as I stated in a previous chapter our skeletal structure resembles theirs.

Human's advanced Intelligence comes from an extra cortex which is not shared by other Species; we also have more cortexes, more brain sectors, more neural networks and a more complicated brain in the range of trillions of cells and neurons, not just billions or less in the case of an ant.

If looked at in terms of Energy and Matter we simply have a more complicated Energy structure than other Creatures developed through Evolution: Animal's vibrate at a different frequency and have less quantity and quality of Energy's for the daily activity: A Human can sustain activity without sleeping for even a week whereas other Creatures have to sleep sooner and more often after sustained Energy outbursts.

One can see the development of a Creature throughout Evolution as Energy Lines crisscrossing, interacting in the body and mind forming Matter. The body and mind consist of intersecting Energy Lines where more or less for each individual couple over each other at the heart, the joints, the organs, the throats, the parts of the brain and other parts of the mind and body, not dissimilar to the Ancient Chakra Point's in Ayurvedic East India and Ancient Acupuncture Point's in China. Other systems in the Modern Western Civilization are starting to make steps in this direction such as Brown Muscles and Bone Integrity in fitness and sports and biological neurology, nicotine and other kick off symptoms in Medical Sciences.

A less evoluted Creature is therefore a Creature with less Energy and less Lines Of Energy interacting, in other words a smaller and less complicated Energy Matrix comprising their being: This determines who you are and what your strong and weak points are. DNA is, of course, closely connected to this Energy Matter Matrix and transferrance process of the Energy of an individual, being in many ways its code.

The Evolution of societies, industries, technologies, computers and the expansion of the Universe can also be perceived in this way: A development to more Energy and more complicated Energy interactions. This is why it is necessary to learn in the course of Evolution; a stagnant depressed individual inactive not learning, only sitting on the couch and eating potato chips, does not progress and worse does not evolve, in fact many are saying these days with the blow hurl suck Hyvescools where all you do is get bored to tears that we are devoluting and dropping each year by 1% in IQ Level . . .

This Theory of Energy Matrix, I just coined it today on 20-02-2014 since I looked it up in Wikipedia and there is no entry yet, is nothing new under the Sun or above the Black Hole, many before since Darwin and Kant with classification of Races and Species and categorical definitions leading to Newton, Planck, Einstein, Heidegger, Schrödinger and many other Mathematical and Physic's number, particle and wave theories which the West since the Pre-Socratic's have tried to put into formulas and axioms which the East has known for millennia since Lao Tse Chung, Mao and the Tirthankaras, which I have proposed on 07-03-2005 has long reaching consequences for all sectors, especially Space Travel because Space Time also follows the Quantum Law's of Point's, Lines and Dimension's of Energy in macrocosmic and microcosmic Energy Matrixes.

This theory is supported by the existence of quanta and q-bits in the Quantum Theory.

This theory can potentially explain the supposed existence of Karma and Reincarnation perceived to be a mere transference of continuance of certain level and complexity of a developed Energy Matrix system i.e. your Soul and/or Spirit entering a new Mind and Body usually via a vessel such as eggs and/or a womb. By the Law of Conservation of Energy this is potentially plausible though no one unfortunately has any proof or evidence to bring back from the After Life yet, if ever . . .

There are look-a-likes in History Of Humanity coincidentally and remarkably also being with Keanu Reeves himself playing Neo in The Matrix trilogy, his painting from 1735 CE is so completely identical to him that everyone thinks its forged and/or Time Travel and/or he really is Immortal; he is also not the only one who have almost identical look-a-likes in previous centuries and even millennia ago . . .

The reason we go Tabla Rasa and cannot remember anything is because your new Host Body does not have the same brain cells, neurons and/or structure so anyone by this fact claiming they remember who they were as some Hollywood wannabe bimbo blonde blow puppet job is obviously a spoof.

Also more radical factions in Theosophy who claim that Blavatsky actually brought back evidence somehow through some form of séance object transference with communication with the Dead Spirit's are like any other close-knit group where you can lift a table with your knee while holding hands in the dark with one candle and moaning and shaking a lot next to crusty Granny nothing more than on the take with each other.

These things do not constitute evidence and most certainly not proof.

The only way we could ever prove the After Life, and such is not needed to prove this Theory of Energy Matrix, and the only way white coat empirical Scientist's (see previous chapter Null Religion) will not laugh it all out ad infinitum is if a Soul or Spirit can be put on a Computer Screen and not like fuzzy white noise and dark wavering shadows of small little darklings suddenly flittering by in the walls, 'Oh my god there is another one, wahhhhhh, they're comin' for meee, wahhh . . . ' SPLAT!! Holy Gruesome, Spy Kill!

The only thing we really know for sure to date to quote their own empirialism is that when someone dies they are A. No longer inanimate, sry bad joke, animate B. Do not have the same metric body and/or head stationary, I mean weight . . . I think I really do like B—Stupid Violent Black Humor, so don't take my Sense Of Humor as an excuse for tolerance either, I mean, I too have heard and seen so many bogus theories, evidences and proofs, uhh proves, which people take so seriously and even form Cult's and Sect's about and then, this tops it for me, commit mass suicide to prove their point . . . SAD.

On the other hand, by argumentation, a GOD who would only give you One Shot At IT could only be the worst kind of cruel anarchistic Chaos God and everything you do was for nothing, has no meaning, do whatever the Hell you want, never come back and off to Nothingness with you, thus how and/or where any Moral's, Ethic's, Good and/or Evil?

This Theory of Energy Matrix, also hand-in-hand with DNA, can explain how through blockages or fluidic streaming of blood Energy one experiences sickness and disease or health; if you have specific weak Energy Lines through certain parts of your mind or body then you will be more susceptible to disease there: Many Leo's have back problems and many Cancer's have stomach problems and if you have neck, spinal and/or bone problems across nerves, veins, muscles and/or tissues like your blown knee cap, heart disease and/or RSI then it's almost already 2-Bit Obvious that your body has Lines, Point's and Dimension's of Energy. Your Energy Lines basically underly and superimpose all of these pathways and also your DNA code and the double spiracle chromosomes at a quantum level.

This theory states the more complex Human brain with more cortexes and sectors probably comes from more Energy and greater Energy complexity resulting in a new layer of the brain since Matter comes from Energy and vice-versa. As to the brain-body size ratio now. According to Scientist's it's also not a question of how large your brain is but the ratio to your body which corresponds to Energy input and output and I am not in violation of, I wonder actually if you could therefore not make a Genius Giant. The Military is also supporting these theories with EM Pulse Blast Weapon's, the Black Humor is you can theoretically murder Granny if you aim a strong enough focussed Null EM Pulse Blast Attack at her head in her fragile venerable condition into her crossword puzzle and finally put her out of her misery, like Anesthesia supporters, the last shocking New's in NL not only a week ago that now Children can request such too . . .

This Theory of Energy Matrix with Group, Individual, Particle and Wave theories also explains effectively the development of Civilization's and the creation of an Information Technology Age at its greater quantity, quality and complexity of Information exchange. In this case, with the invention of the massive super computer Titan, it is therefore not impossible to make a Genius Giant.

Many more examples can be mentioned and thought of to back up, provide evidence and prove Theory of Energy Matrix such as we use only 5-10% of our brain at any given moment.

Evolution, therefore, leads to a larger quantity and quality of Energy and a greater complexity of Energy interaction with Point's, Lines and Dimension's of Energy.

I choose therefore to lean towards a Near-Infinite Continuum of EM Field's and an Infinite Timeline, not only Finite otherwise there can be no GOD, where Energy and Matter transform back and forth regularily and irregularily.

Also, the purpose of Theory of Energy Matrix is to propose this quantity and quality and complexity of Energy interactions is less in Plant's and Animal's and more in Human's and it can be potentially put into a ratio formula of quantity and quality of Energy to complexity of Energy by future Mathematical and Quantum Physic Scientist's who can propose a whole new classification system of categorizing life on Earth throughout Evolution.

As they say 'always leave something to the Specialty Expert's' to mildly paraphrase since ever since Hyvescool I've always sucked at Mathematic's . . . I did follow all of the Beta courses to half way through the last year and since then I've not failed to read and do a lot of Information Technology, Science and Technology, written plenty of Science Fiction/Fantasy, watched to date about 9000+ Film's and Series, played almost all 3D Games except the latest greatest shit which no one can pay for anymore with all their Rich Boy's With Their Rich Toy's but then I dropped them to focus in the Alpha directions of being an Author, Artist, Actor, Musician, Poet and Philosopher.

Good luck! You now how they are with all that resistance . . .

The Voice Of Reason

Regardless of my own argumentations at multiple cross-reference points in my works in FREE Draft's and Final Edit's which are published about One Big Reality = GOD there is still from our points of view as Scientist's, IT Specialist's, Technologist's, Politician's, Economist's, Author's, Artist's, Actor's, Musician's, Poet's, Philosopher's, Celebrity's and many other people in the 21st Century who now form an approximate Strong Majority of 75%+ of the entire Human Species who consider there is still no evidence and/or proof for GOD, always above always looking down, yet ever silent . . .

Refutations of an ONLY ONE Monotheistic GOD: Let it be known, the following points with ONLY the acceptance of the theoretical possibility of an omniscient and omnipotent GOD are the denial of proof and/or evidence of the existence of such a GOD:

Creation after the Big Bang is more likely to be a collective Group effort by People, Leader's, Heros, Demi-God Heros, God's and Goddesses whether real and/or virtual like where we live now i.e. Zeus and the rest of Mythology around the World and self-evident the natural formation of Planet's, Moon's, Solar System's, Shadow System's, Galaxy's, Universes, Dimension's, Heaven Planes, Hell Planes and Time by random and not random collisions, impacts and clustering in Space Time Continuum.

Big Bang Theory, Darwin Theory, Quantum Theory and many other Religion's, Philosophy's and theories all propose alternative possibilities to ONLY Monotheism.

Buddhism, Zen Buddhism, Tao Buddhism, Confucianism, Hinduism and other Eastern Religion's and Philosophy's with theories of Everything and Nothing are all fully acceptable views of Reality's to them and many in the West.

Values derived from Logic, Reason, Rationality, Emotion, Morality and Sun, Light, Truth, Love, Wisdom and Black Hole, Shadow, Lie, Hate, Ignorance and Good, Neutral and/or Evil also state otherwise for Space Time is, as according to Pythagoras in the West and many others, also FOREVER intrinsically bound into Duality, bipolar opposites and Law's of Opposites, Balance and Counterposition's.

Science and Mathematic's with Scientific Method also strongly disagrees for without evidence and/or proof of anything it does not exist except as an abstract theory.

Nature herself, mutations, other Universes, different Timelines, other Planes Of Existence, the absence of Time itself for like my Poetry also inspired by the Pre-Socratic's ' . . . if you count 1,2,3 . . . 4, that is what a clock is for, it gives us Order, Reason, otherwise Time does not exist . . .', and you can always place an Infinite Quantity of dots on a line which must loop around somewhere, thus the symbol: ∞.

The differences between Judaism, Christianity and Islam, respectively, is the 0 and 1 at the same time, ONLY the 1, and the 1 through the 0, thus the different interpretations of the Father Spirit relative to the Mother Matter. I still think it's interesting how the German word is also Vater and Mutter which lines up with Pater in Latin and Matter in English and that One Big Reality can ONLY be a 'him' with Issues of heterosexuality, bisexuality and homosexuality across procreation of Species and Races in Human and Animal Kingdom's which the Monotheim's cannot agree on themselves even, quite vehemently opposed in most cases, all point to the denial of a single omniscient omnipotent GOD.

The absolute fact by reduction ad absurdum of the existence of Alien Species and Races now with 1000's of Galaxy's shown by Maya Spy Sat who can by Evolution only have a different, dissimilar, barely similar or similar Theory of GOD, God's and/ or Goddesses than we do also suggests that there is not one GOD only . . .

I switch some of the orders of words here to show my meaning and what could very well be the different interpretations: Do you mean ONLY ONE or ONE ONLY . . . or do you mean Single with no Girlfriend or Wife, or do you mean 1 is the Power Father Spirit and 0 is the Energy Mother Matter, and this is not per se incorrect since Matter can be transformed into Energy, and some therefore conclude the way to ascend is to transform Spirit into Soul, or as others say you have to free yourself and rise to those realms . . .

Atheism is actually highly logical and not anarchistic whatsoever.

Anarchism, especially these days, is almost borderline Luciferianism since it fully embraces Individualism, Chaos Theory and Free Will. Put into the hands of the wrong person and you have another Terrorist, not that I'm Anti-Anarchism or anything.

Now here is the clincher which is to me kind of like the Monotheism's doing themselves in already with their own Golden Shovel's again: Who defined Satan? Who defined Lucifer? Who defined the Devil? And, my god, who defined her red bikini? If they allow for the Protagonist then they must allow for the Antagonist, if they allow for the Hero then they must allow for the Villain, if they allow for GOD then they have allowed for SATAN; this could very well be one each of ours and theirs critical errors and misinterpretations: SATAN ≠ Lucifer. Proof: Lucifer is a 'Fallen Angel', an Angel is not a God, though is an Immortal. Just as a Demon is not on the same level of the hierarchy as SATAN or the Devil. As I stated in my previous chapter Null Religion, which means Null EM Religion in case you didn't get it, Lucifer is the son of Satan, just as Jezus is the son of GOD, just as the Sun is the opposite force of the Black Hole, Light vs. Shadow.

This affirms my belief that whereas Moses, Jezus Christ and Mohammed was/ is real the Kabbalah, Bible and Quran are like the Bhagavad Gita being epics about primarily Moral's and Ethic's of their culture and history to help give them a Guide and Guiding Hand through Prophet's, or the return of their Prophet who is the Immortal Son of GOD.

Not surprizingly Buddhism with their Tirthankaras has the same epic though they say that the 24th one was the last one for a very long time since our whole Species is entering a Pralaya period since it went through its 14th Manvantaras. If you look up what these mean at Wikipedia then you get shivers through your back for they all line up exactly with our time period in beginning of 21st Century . . . Would it also be on the nose with the numbers 2013 and 2014 one wonders . . .

The childish superstitious view of Heaven and Hell FOREVER when according to their own Award and Punishment systems in their own paragraphs is another misintrepretation by many, ' . . . it's more like you'll be going away for a very long time and it only feels like forever . . . for how many centuries of punishment to the Hell's would you give a Mass Murderer and how many at a min of do you have to kill before it's no longer just a Serial Killer and how long do you have to be a Serial Worker before you are liberated . . . ' Instead of an Absolute Real Existence which everyone is subject to globally, an exaggeration of such Karmic bindings, punishments, dissolutions, solutions and resolutions the mere subjectivity of your own very limited microscopic human beliefs are indicative of the fallacy of such Judgement by ONLY ONE GOD: I actually believe and follow myself primarily in a combination of Celtic Baptist Christianity with a number of other Pantheon's and primarily Zen Buddhism, so am I also not judged and sentenced by my own GOD, God's and Goddesses, anything else is like being kidnapped and/or enslaved again in a Viking Iron Neck Collar.

Karma, if it can be proven to be actual Quantum particles with binding of Soul, Spirit, Mind, Body, Energy and Matter, thus also once again on a Computer Screen and repeated by Scientific Method, is also denial of some omniscient and omnipotent GOD receiving an ad infinitum quantity and quality of email messages i.e. prayers and hope that the whole Planet Earth is not descending into Chaos And Destruction, at the same time, since there could also be an inherent Natural Mechanism governing entire Reality though with the influx and influence of Soul's and Spirit's and their Will's and Free Will's.

Saying it's your Right Of Choice Of Religion And Belief is fine for Strength Of Faith through hard times, however is purely subjective and Blind Faith does not put bread on your table and does not prove anything, Blind Order's get you only killed like in WWI which also only led to WWII anyway.

Now onto the highly repeated attempt at a lame retort from primarily Noobies who went confronted with these Topic's for some reason first get highly emotional and then only after much Debate and/or Battle screw their thinking cap on: It is my choice to choose whatevel code and/or belief I want. Well, on the one hand in Modern Western Civilization you do not lack that right, however on the other hand it comes across as the worst Type Of Noobie who has no more than the equivalent of 1st Year Of Hyvescool in EU . . . Karma is also not some Stupid Human taken Justice into his or her own hands.

No omniscient and omnipotent GOD would in the First Instances and Last Instances ever allow such Evil's and atrocities in History of Humanity to take place and GOD does not play dice, however SATAN and Lucifer have no problemo with such, and Free Will, the best and worst gift of GOD, allows you to choose between Good, Neutral and/or Evil.

Esoteric systems are also blatantly Anti-Monotheistic and also have many Good Values such as Wisdom, Truth and plenty of theories to present to Modern Humanity which is parallel to this argument.

Jezus Christ and/or GOD at Times Square is a good stunt! We're still waiting . . . not to mention how many more Doom's Day Prophecy's on Internet? I, myself, am primarily a Realist and sometimes lightly optimistic though I have been refined and sharpened with a Silver Tongue and now and then a Black Tongue through Life Experiences and some really stupid and nasty people along the way. Basically, I think as many that we are not ungrateful to be finall out of 2.5 millennia of Monotheism domination by all three and have finally hit the Information Technology Age with unlimited Multimedia's and Right's Of Choice though primarily still only in Modern Western Civilization though a lot of problems remain unchanged such as differences between rich and poor, Power And Energy structures, imbalances, faillisements and some ask as to whether we have simply, once again, replaced the Chimera for the Hydra.

Order in the Chaos, because GOD has 'more than 2 hands' and Jezus is somehow 'managing' all of it right now is a little bit hard to picture though he does have his Henchmen for if the Waiting Row to Hell and everyone's got their own Ticket Number is really really long like they're all trying to get into it rather than run screaming away in terror than the Waiting List to Heaven must be even more mind boggling: The denial of intrinsic automatic Natural Mechanism's in the Universe is absurd, not only in denial of the Scientific Method but Science, itself, and last we checked it's only really a small fraction of fringe Group's who still purport that for how do you even get by a single day anymore without using at least a mobile? The world is much smaller now.

The Monotheistic Religion's and belief system are by definition circular in logic, self-fulfilling in Prophecy and predeterministic in Philosophy. These qualities are flawed as argued by countless others already, especially many renowned Philosopher's. To argue ONLY ONE GOD all the time is a Moot Point and not even a retort.

The Holy Bible, The Bible or Bible, this was fun to look up at Wikipedia, I mean talking about Object Name confusions, the best sold book in History Of Humanity was mistranslated a number of times and like the other Monotheism's is misintrepreted all the time which unfortunately not only leads to conflictious Debates but bloody Battles, and now we're seeing it recur again with them in 21st Century of the likes which has never been seen since WWII.

Constantine's Holy Vision of the Holy Cross, over a blood battlefield, is fake and led to an incorrect binding of the Church and State, which is what they always wanted and was inevitable and led to +/-1700 years of persecution, repression and suppression of all those opposed to them as heretics to be interrogated, tortured, burned, killed, murdered and/or sent off to the New World. Such symbolic stories can be interpreted in very many ways but historical facts are less erroneous unless of course you want Queen Isabelle to invent Christopher Columbus again who was more like the last to discover the New World, even Vikings made it there sooner and pre-Homo-Sapien's walked and boated via Russia, Bering Straight, Canada, America and South America to populate it. And how exactly is a battlefield spiritual again? No, it's to wipe out their Enemy's, all defined as Devil worshipping heretics and pagans which is an Insult not just to Celt's but to many other Religion's and Philosophy's.

Don't get me wrong, which I have been falsely accused of myself for so many times, years and decades thanks to my Parent's Anti-Sentiment and the previous Vatican Right Wing Regime who falsely prosecuted me excessively and for too long for simply being another Celebrity who had a Beer and Marijuana Habit for 1.5 decades, I am not Anti-Orthodox, it does not lack its fraction in my system but not in that lack of Light of Christos or in those factions who desire ONLY Power And Energy since WWI and before. My own Lans Family fought against them as is proven by the Lans Admiral raised to Royalty in our side of Germany and Netherland's. I've never been Right Wing and never will be and it's just phenomenal how the whole Planet Earth is now Reseller Galaxy with 2 Genius Giant's about to square for the final round of show down boxing and club swinging while everyone else dodges, dives and ducks for cover. QUACK!

Next to other misinterpretations and mistranslations, 'communication errors are the cause of all conflicts', there are very poor and popularized modifications of Greek and Latin words through all Medias just like everyone thinks Hades = Hell, no they're not the same God even.

Omniscient: The word 'omni-science' means much less to the Greeks, meaning more 'versed in all the arts' than actually 'all-knowing'.

Omnipotent: Likewise, 'omni-potent' means actually an all-round highly skilled virile individual.

Both words are therefore completely blown out of proportion in modern interpretation.

The acceptance of the theoretical possibility of a not all-powerful, not all-knowing GOD, regardless if is the same as the Creator, is still denial of the evidence and/or proof of the existence of an all-knowing, all-powerful GOD.

Due to the Nature of Near-Infinite Reality as argued by Science, Mathematic's, Religion's and Philosophy's through multiple means, see also my other chapters and works, and the dubiousity of an actual existing Infinity, except for an Infinite Timeline otherwise there can be no Infinite GOD, as argued by many too, the chance of an Infinite GOD is nihil and absolute for ONLY Nothing can be Infinite; the very nanosecond you enter Finite and/or Near-Infinite Space Time it is no longer Infinite.

Therefore, Time = 0. Or, Time = ∞. They have not figure out Pi yet, in fact the most powerful computer on Planet Earth has still not figured it out. Or, Time = whatever the Heaven you want it to be and throw it in or out of their equations.

'The only way not to make any new Karma is to not exist.'

Regardless, of the different interpretrations, such remains a Paradox.

Hopefully not a Pandora's Box or did they already open it: ' . . . and all the plagues, sicknesses, diseases and evils shall be released upon the Earth leading to the Hell's . . . '

Jezus Christ is a God, the Son of God, and is accepted as such by the vast majority of everyone on Planet Earth who is capable of miraculous healing and resurrection and if he will be coming back also reincarnation, unless he never died . . .

They having then or now trying to talk to GOD, GOD returns only silence.

Thus, as stated as the premise of this essay there is still no evidence or proof, real, existing, historical, scientific, legal, investigatory and/or by location whatsoever leading to the evidence or proof of the precept of an Infinite GOD. There is only the grave of Jezus Christ which was found with the names of his relatives which was calculated at being at a chance of about 1 in 8.7 million which is highly debated since such is technically only circumstantial without enough corroborative evidence. Read 'Het Graf van Jezus' door Jacob Slavenburg who's book 'De Hermetische Schakel' I helped translate and did the final edit to English called The Hermetic Link. He also helped with someone else to translate the Nag Hammadi scriptures.

To conclude this chapter: The question here, is not whether GOD exists or not; the argument of GOD is irrefutable, there can only be ONE BIG REALITY which was created at some point. The question here, and is the purpose of this chapter, is who, what, where, when, why and/or how is GOD? Only when we answer these questions about GOD correctly will we understand the Universe better, if not the Multi-Verse . . .

Classical Newtonian Physics vs Modern Quantum Mechanics

I am not the I but if I am not me then who am I?

It is proposed in modern terminology that there exists a Virtual Reality with and/or apart of a physical Reality.

The difficulty, of course, is Human is primarily confined to the senses with devices and extension of such. What classical Physic's has taught us is the hope there is some Absolute Reality with absolute Law's. Quantum Mechanic's is presently contradicting this conception and allowing for the belief in i.e. Intuition, knowing something is coming around the corner and you step out of the way just in time as the car comes blasting by . . . Also as I have brought up on a number of occasions with Family and Friend's what is the difference between a significant and insignificant Event in your Timeline; can I just turn left instead of right on a day and my entire Timeline changes or is that simply meaningless . . . Some things are obvious: Do not go into that street or neighbourhood at 03:00 at night or you'll probably get stabbed, kidnapped and/or hijacked. Other things however are inexplicable and can only be satisfied by the Chaos Theory and/or Karma Theory for at only 24 years old after one week of work she died horrifically in a car accident, what was her whole Life for, her Family, Friend's, school, post-education and only one week of work, it seems the worst injustice even and that GOD is only the worst Evil Dictator again, well no, I keep saying that you keep forgetting SATAN or Lucifer . . .

Quantum Mechanic's is not only allowing for such unexplained phenomena, lacking evidence and proof in physical Sciences, but is breaking the classical cause and effect, Space And Time, physical structural frameworks; something now does not have to necessarily be caused in the past only for as some like to say Time is not linear which I consider a little bit of an ambiguous term since there are Timelines, unforturnately to allow for Infinity they could be a little loopy; I am also stuck in some kind of Time Loop here in NL for I cannot seem to escape from a certain amount of things which keep recurring in my Life next to those god awful Catch 22 Scenario's . . . Hindsight is always 24/20 but maybe I really should not have wasted a priority signature to move to a B—Motel with no insulations of any kind to escape living with my Mother since the advertisement had a spelling mistake on it: 22nd floor, no lift and I was the only one who reacted to it: Just to bury the hatchet with the neighbors at my previous address forever, if you look at it this way, it's not really anyones fault cause all you have to is fart again and it echos through all the Near-Infinite compressed spaces of Time and realities . . .

If there is one large/small now and here then wave particle causation with or without the Observer and some kind of Timeless Space is possible. The 12th Monkey Effect which is the Object Name I like to give it though it should be 13th Monkey Effect has been demonstrated in Evolution countless times over, not to poke too much fun at the Chinese again but now they apparently can just rip off everyone's patents these days via Internet just like Japan did to date and also blame everyone else at the same time. Everyone, even the most macho of man, has Intuition yet no one can prove it. This is an argument I have not failed to use on a number of occasions, just like everyone gets affected by different Types Of Full Moon's in different ways no one can prove it again . . .

The problem here is the actual Methodology of proving something in Modern Science to date which relies much to heavily and only on the Scientific Method which was one of their breakthroughs, one can also say just remove Time from Einstein's equation and we are done with the whole Debate again but that is of course a little bit blatant and not a Good Methodology. However, what is a Valid Methodology and who gets too, again, decide what is the Official Methodology and define the point system. As far as the Vast Majority is concerned if everyone suffers or enjoys it like the Budget Crisis or food and/or drink of different types then it is already proof yet no one has any evidence now . . .

How do we really give evidence and/or prove anything these days without it being shown multiple times, not just 3 times even, to thousands of White Coat Expert's around entire Planet Earth on a computer screen . . . Yet my practicing of my B—Stupid Violent Black Humor Live Session's for my next Vid Humor which will simply be called Alien's Lookin' In 02 in my own living room in The Free Show not because I'm doing it for you as a lame lonely suck nerdy type has to first go through the Medium in each and every First and Last Instances: Numbers can be toyed with, interpreted incorrectly and if you have even only 01 FPS cut frame or a slightly different angle of only 01 centimeter then it not only excludes such from the entire reality but changes the meaning completely of their entire realities; also really obnoxious is ripping 'non-'s' or 'not's' into Audio Channel's since you are acting and/or roleplaying like an irritating Nüber Püber, I say if you rip it into their sentence at the wrong moment even across Cross Channel Interference of their Island Mentality's that you make another National Insult and could cause another Civil War even and not just mildly piss them off . . .

It is a basic self-contradiction to state one can prove something in Virtual Reality with virtual Rules and Law's using physical Rules and Law's, which is why each of your systems and Internet is suffering both ways and causing Hydra complexes of the worst kind, using a physical Methodology for a Virtual Model, called the Scientific Method which comes from classical Newtonian Physic's though no one still really knows how they figured it out back then since when a physical Object falls it still encounters all kinds of interference and thus never really hits that mean speed except in an Absolute Vacuum; did he stick his hand into an Absolute Vacuum trick again . . . To allow for other Virtual Reality's which interact apart and/or seperate from the physical Reality one must concede an equally viable Virtual Methodology of discerning and proving Virtual Phenomena, thus even Paranormal Phenomena since such by definition is virtual, spiritual, mystical, psychic, psionic, mental, emotional and/or immaterial, thus without a doubt NOT physical. One point of note though which East Indian's like to poke fun at us is technically speaking anything in Reality is also physical having some Form of body and/or vessel but for lack of better terminology and to distinguish between that which is visible and invisible, material and immaterial, the words 'real', 'physical' and 'virtual' should be clear enough for the Saki of the Argument.

It is not possible to prove Virtual Phenomena with physical Evidence, physical Rules and Law's and/or physical Methodology, therefore with only the Scientific Method. I am not saying toss out the Scientific Method which we could still utilize in some ways since we still have to repeat the evidences and proof without it being really Great Art, Special Effect's and Radio Ear Devices again which is basically how I have no problme disqualifying ALL Paranormal Experiment's to date and, once again, for all of the really Great Hollywood Stunt's there is no one on Planet Earth who has levitated without a Technological Device of some kind to 200 meters at Times Square with plenty of witnesses; even then it would still have to go through the Medias and Medium and could also be disqualified as another Mass Halluci-Nation.

This, I imagine, will cause a mild uproar, but Quantum Mechanic's is actually purporting exactly such for we are now seeing that not only the Quantum Mathematic's are showing such but the actual Quantum Physic's are too i.e. NOT Local Causality of two particles at a great distance simultaneously.

There is no need for this uproar and/or protest, since what I am suggesting does not actually delete the Theory of the Scientific Model but rather provides a new framework from which actual progress can be made into Paranormal Field's, no pun intended, other Reality's, other Planes Of Existence, Dimension's and other inexplicable virtual phenomena such as the previously mentioned so-called and perceived Random Event's.

In fact, in the long term this proposal can be used and expanded upon to prove the above Field's Of Specialty's, and much more. What I suggest is a logical reductionary system Real and/or Virtual which can be used within and/or without the existing Classical Newtonian Physics vs Modern Quantum Mechanics so we may actually be given the chance to provide evidence and prove Virtual Phenomenas and Reality's which everyone knows they have but cannot stick their finger on it.

I call this Virtual Reductionary Relational Matrix Model Theory.

This is based on the fact everything existing and/or non-existing can be proven, no matter how difficult or how much quantity and quality of Time it takes, by the usage of logical reductionary systems not lacking evidence and proof at some point in development.

Also a good example is in sports and 3D Racing Games which I call Lines Of Energy but could also be called Lines Of Force or even Lines Of Motion through dynamic interaction of 3D trajectories and momentum. What is actually deciding whether you wipe out or not in my Numero Uno Favorite 3D Racing Game called NFS are the invisible lines and variable which surround an object as it blast by at 300+ km/hr equivalence. What decides you have to ask yourself when you just barely hit the corner of the building as to whether you spin and bounce left or right or slide along the wall and cause lots more free fun and gratuitous Property Damage and Car Damage . . . There is always more where that came from . . . This is not perceived by the Sport, Fitness and even Olympic World as paranormal yet in even previous decades only we were completely oblivious to these underlying forces and study of such has led to new Track and World Record's . . . Are we not actually confusing a lot of so-called Supernatural, Paranormal or Virtual Phenomena with simply already existing though invisible Quantum Phenomena . . .

I also like the following example which is also in my Spy Kill's 02 though phrased differently: 'Can you prove your Not Unreal Identity = Real Identity in the thing: What is your Not Unreal Name = Real Name, Address, IP Address, Frequency and License in it? No? You therefore do Not Exist in it and get Jack Shit Nothing. Holy Gruesome, Spy Kill's! Like I have said in previous chapters, I like a Sense Of Humor to kill the dryness.

Thanks though, again, for all the FREE Entertainment! (This statement is, of course, inspired by Hitchhiker's Guide To The Galaxy when next to clear out Planet Earth for an Interstellar Highway I also like their statement ' . . . and thanks for all the free fish!'

'Where does something hide the best? Right out in the open.'

'No matter how absurd something may seem, if reduced to the only possibility it must be true . . . '

'It is only a matter of time . . . '

Change

This Society is not going to self-destruct like so many other Civilization's in History Of Humanity. It will merely change: 'Death does not exist, it is only change, transformation and transition to a new face, new form, new fashion and a second chance to dance . . . '

Nicht Ontheil Chronos!

And we are not talking about per se huge change either: Change, fluctuation, adjustment, reevaluation, improvement, increased living style for all. As the population increases this will be forced or there will be a drastic population decrease due to famines, sicknesses and diseases, wars and/or chaos. These, however, are very unlikely in the present state and growth of Modern Western Civilization. However, on the other hand, which we now see happening again, the other Colony Country's or just bloody well everything near and below the expanding heat bandwith of the equatorial belt are *very* susceptible to such.

Change is inevitable. In fact, there is nothing but change. Each and everything, each and everyone, are in continual states of fluctuation, not just your blood pressure either . . . If we would finally just scan people more in Hellth Industry and other sectors instead of performing all of these now outdate overly expensive procedures of, once again, a whole array of causes and effects of symptoms, syndromes, practices, cures, treatments from the whole blow hurl suck process from step AA → ZZ which for some is not even reachable as you fail at Step 296 in Anger Management Course or still suffer the worst Insomnia leading to No Serotonin Surreality then we would see all of the spectrums, thus quite literally the sub-spectrum and potentially smaller than nanoscopic already . . .

People fear change: The introduction of the Unknown. The very word makes you think of a dark forest with evil rogues and beasts . . .

Notice how strong Habit And Tradition holds, 'Your Habit's and Tradition's, I'm not judging you or pointing my finger, my HAT off to that, slash through your throat, a little tap of my cane, Holy Gruesome, Spy Kill! What each and everyone one of us do, the most horrendous conflicts we cause in each of our relationships, just to maintain our own Habit And Tradition. Tradition is comfortable. Habit is tasty. Change next to bringing much needed improvement in multiple sectors is also the elimination of Habit And Tradition. However, there cannot be any long term fixated static outdate holding and maintaining of ANY Habit And Tradition: Change is the only constant and you will be doomed at some point by Law's Of Physic's even to change even if that is 'Until Death do us part . . . ' Nicht Ontheil Pluto, Nicht Ontheil Osiris, Nicht Ontheil God Of Death.

In fact, you can see how change would be desirable if it leads to real improvement which does not lead to only Better Coffee as it is forced to introduce The Unknown. This one now sounds like a Great B—Horror Thriller Suspense Movie Title . . . which I guess if we are doing that 2D/3D Object Game all the time, did I accidentally just copyright for myself since you cannot touch My Unique Combination or is it always just first come first server on Internet: With more variables, greater solutions can be reached though if change is too fast all at the same time which is unfortunately the case on Planet Earth from start of Internet then great calamities, disturbances, wrenches and Bottleneck Effect's with a drastic increase in Chaos Factor's will not fail to increase and dominate.

Nothing is stationary. All particles are also in a state of fluctuation. There is, in fact, not a single object in entire existence which is not in motion, except for the mythical and also still never found Dead Center of the Universe from the Big Bang Theory. That is now sounding like My Joke about The Dead Centre Shopping Mall where Vampire Demon's and Gothic's of all types hang out and still get to smoke inside standing in strange angled half-broken poses while buying everything from the cheapest second hand clothing to the most expensive chique designer modes. However, the opposite theory states that is actually the Throne Of God, well no, not really, if the Galaxy does not fail to collapse back down and be swallowed whole, no pun intended again, by Black Hole then I just don't think it's too goody woody twoo bwoo bwood . . .

To respond to the changing environment it is also necessary to adjust yourself which is the hardest part, I mean do you know anyone who is not a Rammy Bammy in some way or another and when confronted with their own Issues put up a wall which makes the Russian Iron Curtain Effect look more like its upgraded model to Russian Iron Steel Titanium Wall Effect. This is the 'Survival of the Fittest' still commonly misunderstood as the 'Strongest' by manly when you look at Hyvescool behavior and many others.

Adjustment can also be the balancing of Energy's. All things to equilibrium, balance and even rest through the excited state of LASER's, release or expenditure of a heightened state of particle to wave or beam and then back down again to less agitation, like one of those thin always nervous agitated nerdy types who does not know he/she has Hyper-Hypo-Allergic Reaction's to almost everything and keeps knocking back all that white sugar and flour cause they are also usually the worst Diet Noob's. This is quite literally in Law's Of Physic's supported by Newton's Law and Law Of Schrödinger plus too many others to list here (go to Wikipedia even for all other hidden and/or 2-Bit Obvious references) for the purpose of stability, balance and equilibrium: 'All things to Balance'. It is a Law of Nature which decides the course of entire systems, Planet's and also helped form the contents of Galaxy's after Big Bang through impacts, collisions and then orbits of different kinds. And anyway, where else are we, but in Nature? I've always found it somewhat of an oxymoron to state

'Super-Natural' and like never 'Sub-Natural'. It is just all in Nature if you equivalate Reality = Nature.

Example: A force comes into being. So as not to go spinning out of control, we adjust our corresponding motion and position. We learn new ways of surviving and ultimately progressing, we adapt.

In the process of this re-evaluation an improvement can be reached, you learn how to survive and progress better. It is this simple. The problem is, of course, there are individuals who give resistance to each and every idea regardless of how correct and reasonable it may be or not. This is also what I mean by some Anti Anti Anti Noob in a previous chapter who just shows up to irritate everyone with Only Semantic Style.

There is only one obstruction and this is: It is within Human Capability to make the Right Of Choice Of Death, though it is still phenomenally difficult to commit suicide even if you are on Drug's And Alchohol up to the hilt in your neck and still suffering the worst Trauma from being any Type Of Victim. A Human can ignore the incoming messages, the clear, obvious necessity for change. After all, we have Free Will, more or less . . . I say more or less cause I do not dislike or disagree with an Anti Free Will Theory stating like my own Somewhat Not Neutral Administrator that if he/she/they are paying your Salary which pays for your Wife and Children you can only be loyal to that instance, not disallowing the multiple instances.

If change is accepted and assisted then more and better can be realized but it can also blow up in your faces and not just in the Toilet Humor. More and better is a basic thing necessary for Human Evolution. If you do not have more and better stimuli then you cannot learn as you do not have sufficient information to work with and you toss your piece of shit 16-32 bit laptop over your balcony and into the backyard, like an Aunt of mine actually did about a month ago in Jan 2014 . . . You must subject yourself to greater and greater Unknown to increase your Known . . . This, in case you haven't read it already, I have used in Apotheum Colluseum—The Ultimate InterActive™ Game—2nd Edition which is primarily the Battle System of The Free Show and Spy Kill's 02. So, you know it, I'm not only dissin' all your Hyvescool's still relatively stuck in the Colonial Ages compared to northern EU standards but I'm laughing them all out completely: What the Hell do you expect if you're from Step 01 giving your next generation the wrong, poor and bad stimuli . . .

You must Observe and Experience more. See also a later chapter on a full system of this Observation and Experience methodology.

Since most Human beings, and then just not being in your Host Body anymore, have the capability to and will choose for survival and embetterment, as opposed to death, just observe the increasing over-population explosions everywhere under the worst poverty and pollution experiences, the result will be an increased life style for some and famines, diseases, wars and/or chaos for the vast majority: 'You're caught in a prism of your own design.'

'How you define your system is also how you suffer from it.'

'What goes around comes around.'

'You reap the seeds of deceit which you sow.'

'Each and every action has an equal and opposite reaction.'

'Debts not paid in this one are doomed to paid in the next one.'

'The only constant is change.'

' . . . and all I've got now is short-change . . . '

'To them it is nothing but change.'

In all times we are forced to survive, though such is almost a basic self-contradiction. With increased population, we are forced even more to change, fluctuate, adjust, re-evaluate, improve, increased life style for some and famines, diseases, wars and/or chaos for the rest, such is the point of view of Realism on Planet Earth.

This is the way it is, this is how it has always been, it will never change . . . No, error, that is where I now draw the line, that is just blatant incorrect Pessimism of the worst kind which is always 10X Worse than highly exaggerated Optimism and causes real damage. I can just as easily argue that in about 100 years after this blow suck hurl Fossil Fuel Age ends that by about 2100 we will have not failed to Hyper Modern our entire Planet Earth into a Type of Democracy with no poverty and pollution reaching a so-called 3rd or 4th Renaissance with Space Travel and Colony Planet's instead of being stuck in New World Order at 30 years after Brave New World signed in by everyone on Jan 2014.

Since these people in themselves lead to greater chances of survival, pending choice between Life or Death, and they lead to more embetterment, we thereby will be forced to progress through increased population, though suffer some periods of dippy deluges. It is also a Law and a statistical fact, our planetary population is already 10X that date . . . With Resources gotten from other Planet's even we cannot fail, with better crop production through self-sustainability we cannot fail, with another Spy Sky Pan across NL alone again, just take a plane in too, we see it's all flat and empty full of open fields where we could Animal Appartment's (see Planes Of Existence—1st Edition—Published by AuthorHouse UK) and decentraliz out of over-populated inflated Rip Off EU Zone City's . . .

Thus, in contradiction to the standard belief that an increased population leads to all evils, I would argue we actually have to increase our base population, our breeding pool, for what is the ratio of geniusses to normal IQ necessary to colonize other Planet's and develop more Expert's in each Sector Specialty; for each Planet at a min of an elite group of Elite Scientist's and Artist's of Genius IQ are needed to populate the new planet, allowing for necessary weak strands to maintain bio-diversity.

QED.

QER.

End Debate.

Start Betting at 9 and 10:1.

KABOOM!

One With Nature

These are the premises of One With Nature:

1. Nature is a Sphere, a continuum of flowing Energy.
2. Nature has a solution for each and every problem and such is true.
3. We can know and practice all Truth through Observation and Experience.
4. Human's natural Evolution is to become One With Nature, to become a sphere of Energy.

These will be shown to be true, within and without . . .

One With Nature

And dash, we, into the fields of thyme,
Me, four two one, walker in Time . . .
Who was, he who walks over stone,
Who was, Lord of Fusion.
Laughingly gleefully, I trance, with
She who knows all things . . .
Who was, She of None and All Faces.
We kiss momentarily
In the shadows of a great oak Tree . . .
Then run off the Sun's playful dances,
Returning to our future, peaceful Celtic havens,
Will we where, build, meld, blow, and fuse further creations,
Perfect in this order,
One With Nature.

Sustainable Development

Our Planet Earth has an infinite potential to feed us—at least as long as the Sun lasts.

If each couple is allowed at a min of 1 girl and 1 boy per generation then our population will peak out at its present level of development. The fact that various Country's like India, not just China, have no birth control laws and/or removed them is just simply irresponsible, why not just Pac Man the whole Planet as One Big MUNCH! goes through the whole thing each day, or play the Population Game and take over the whole world in no time flat . . .

Taking into consideration all habitable land, this excludes about ½ of Russia, ½ of Canada, pretty much forget everything near the two Poles Arctic and Antarctic, and it's almost becoming borderline in the expanding equatorial heat bandwith region despite people's bizarre desire to live in the hottest, sweatiest and dirtiest climate zones, excluding Advanced Technology which now provides a strong new theory of so-called habitable Colony Planet's, there wil be a 444:1 population to square kilometer distribution, globally, in no time flat, like a decade or two even . . . Netherlands has a 1000:1 ratio already and presently growing a little bit though it has one of the worst procreation rates ala 1.2% only in 2012 . . . however, the influx of Immigrant's is mind blowing: Apparently according to statistics 156,000 people leave Netherlands per year and more come in . . . It seems more like someone added a zero by accident again as a long standing tradition of bad jokes about such things in Netherlands . . . But what is definitely a fact is that that part of India did not fail to increase its population 15:1 from 1 million to 15 million with all their very not unhappy Grandfather's walking around now . . .

One must think vertically and not horizontally in terms of population, if not in everything else which is also peaking out statistics across the board these days through the stratosphere and sub-stratospheres. The numbers alone I have seen myself almost everyday in my smartphone are mind numbing . . . In fact, the way we figure it, next to all the blown regions, Internet is even about to tentuple . . .

We better as well face it: Human will dominate the entire world. We will have City's and Community's, with very many people, on every square kilometer of the Planet.

The wild will not exist, anymore. The wild is too wild, it was a part of our Evolution, where who ate who ruled things and we did nothing but fear the night in the dark forest weir predator wolves rip roar shred and dismember cute little bunny rabbits for a night snack when they oopsy wake up by accident. I also like my joke about Survivalist's who get a lot of fun poked at them these days about no more oxygen and so forth, however that will not be the case, there will be just no more Wildlife, "Hi! Are you my lunch?" STAB into bottom left rib cage with Bounty Knife and slash straight right through, blood exploding everywhere, and your previous Ally Friend onto the BBQ. Holy Gruesome, Spy Kill! And, it still remains just unretortable: A Wild Animal is not a Domestic Animal.

Trees and Plant's will obviously always exist in our Community's, though NL can be quite a Swamp and Jungle now and then because it is technically a sub-tropical region, except in some bad winters, as they do in City's today. If we do that Spy Scan Pan thing again we see that, 'Wow, holy shit, the entire NL is flat and empty except for the City's! What a bunch of Bullshit Politician's and Economist's again all ripping us off so badly in the centralized City Zones of this Rip Off Euro Zone . . . In the near future there will be a definite decentralization of the population as the only thing you need now is a relatively good Internet connection to help her keep shopping and shopping and shopping . . .

Trees, Plant's and Animal's will co-exist with Human's just as squirrels, birds, dogs, cats, fish, insects, mammals, amphibians exist in City's today. Picture, however, a couple things though which is drastically different then most Areas and Region's, except for huge City's like New York: A. Flat's, Appartment's and Sky Scraper's from west to east coasts, thus you might as well start planting parks on top of buildings and hope no one gets blown away by too much wind again. B. Animal Appartment's though not next to Granny's Appartment cause it could be a sound disturbance (see Planes Of Existence—1st Edition—Published by AuthorHouse UK).

Humans will co-exist, peacefully, mostly, barely, though this will take a lot longer in terms of decades to half way through the 21st Century as they all finally figure out two crucial things: A. It's actually Specieism and Millenniaism, not Racism and Centurionism, in other words, don't mind the Black Humor and purposeful misspellinggs, we are actually really just brothers and sisters of each other through DNA of Human Species and definitely suffering the typhus Relational Argumentation's of such. B. Thinks vertically not only horizontally, we can also not fail to make Plant Appartment's, thus not only put a plant next to someone's head while they sleep.

Thus, actually, I just bumped off my own retort in a previous chapter which states we need to colonize other Planet's to reach the Infinite Resource Level which some think is just impossible, a strange delusional myth by lefty greeny potheads . . .

All systems for the maintenance of Human survival and development will be in Human hands, not left to the whims of another storm to come along and blow your whole Village clean off the map or die of the worst Wraith-Like Plague Pestilence. There is, of course, one exception: If Alien's show up then we might not have even that control over own systems for long, by reductio ad absurdum through all the Galaxy's, thanks to Spy Sat Maya again, what are the chances that there is no civilized Life out there and what are the chances over all thousands of millennia since the Big Bang that they are more advanced then we are? Yet, we still have not a sinle Laser Weapon on any Spy Sat or Space Station's, I mean why not just embrace 'em all and let 'em all in again . . .

We *will* control all systems, we have to, it's mandatory for our own survival, that's the whole reason A = A the entire Planet Earth is presently suffering the worst effects since WWII and Chaos is dominating, there is too much in-fighting and we don't have enough self-control over our systems in each sector due to multiple simultaneous Cascade Effect's and Spiral Out Of Control Effect's with massive imbalances.

Would you have it any other way? Right now, things are pretty out of control, if I had to reference all the New's Item's alone it would at today 27-02-2014 require 10's of thousands of pages . . .

Since the majority of us choose Life over Death by the strongest instinct of them all called the very not unreal, though in some cases less proven, Self-Survival Instinct, since 'Survival Instinct' only comes across as 'I don't think, dear Socrates, that they have a problem with just slaughtering the neighbours' and since increased population will lead to increased needs and wants for survival and development, we *will* change our ways.

Even if some persist in their very stubborn outdate Habit's And Tradition's, like their Old School swears on it, by numbers alone we will not fail as proven by Africa which due to their virility despite all of such catastrophes and wars on a not irregular basis, talking about spiky Market's, they keep getting bigger, stronger, faster and better.

In fact, this is the period where such is occuring. Right now, we have all of the knowledge of all of the past at the touch of a button and more is coming. Our Core Group's in our IT Community's each do not lack their Near-Photographic Memory, but Internet itself does not lack an Absolute Photographic Memory, though some things are indeed not on Internet yet, they will be in the Near-Future, and I can now look up approximately 99.999999 of all Human Knowledge, allowing for delay in cross-language and cross-compatibility delays, in less than 12 minutes flat, unless, of course, my piece of shit outdate PC due to technical difficulties is, "Houston, my PC is lagging very badly and do I have permission to hurl it out of the Space Station Module . . ."

At the touch of a button we can call up the Information of all of Humanity, as we know it . . . though I would definitely like to add since we hear it so often in the Medium's that Information, Knowledge and Data is one thing, Internet is the best and worst of all worlds, but to very erroneously presume that it is Wisdom, Moral's and Ethic's is another story completely, we may have all of the Information and IQ Level to use it but no one is applying it in the correct fashion and only sticking it into their own wallet again . . . The potential for abuse to up your own Power And Energy is in fact never been worse.

'If you are wondering where the Hell Culture went to, well, all the great minds are divulging themselves of say, Egypt today, and well, how about Paganism tomorrow, and, uh, I think I'll check out an Eastern Religion on the weekend, oh yah, where will

I fit in Paleography? Gee, I don't even have time for 3D Games, wow, like, 'Video Games' does not even exist anymore which are now Arcade Games. And then they go to cool parties with lots of European friends and share the info, which, tja, sip, glug, puff, smoke, I have to admit even to my Social Media Friend's, I'm a little jealous of . . . ' The greatest source of Information, the greatest amount of Unknown, ready to become Known, the greatest revealing of Reality and Truth, ever, exists here today. Unfortunately, of course, just as with many in the past, the first thing they will do with Laser Military is make another Weapon Of Mass Destruction, only much later will they use it for Near-Infinite Defenses and take seriously for once that Alien's could very well show up, and not for lunch unless it's you on a BBQ, in the Near-Future i.e. 2056 for a blow suck hurl date.

Do I have to scream it? Go and Educate yourself! You are sinning if you do not! To quote again, 'The only sin is Ignorance.' Do you even know who I'm quoting?

This is the time period where we can live in Harmony With Nature. We have come to a state now where it is actually possible to live off of the basic forces of Nature: Earth, Water, Air, Fire, Ether, Shadow, Light, Form, Spirit and GOD. These are the Top 10 Element's of Nature in that order which you could, indeed, say interact all the time and/or do not lack in being the constituents of Matter and Energy. For the first time, with, once again the exceptions of regions like Philippines being blown completely away by multiple consecutive Super Hurricanes, we are no longer only Victim's to the forces of Nature, now we do not lack the Science And Technology to harnass and control it.

We do not need to grovel in the dirt. We do not need to shovel shit. Now there are these totally Massive Monster Machines to get crude oil in Alberta and its apparently unlimited except for the off-the-scale consumption level of America at 2 million barrels of oil per day. They are actually not the Evil Bad Guy who is no different than anyone else supplying and providing a need or want in our Bust Boom and Supply and Demand system, in this case it's actually the Dumb and/or Stupid Consumer Effect, thus sry there but if your people actually consume that quantity and quality then don't blame MS again!

It is everyone's responsibility sitting in this comfort to support this State Of Comfort—this is how to adust the system to allow all beings on Planet Earth to have this comfort which is also a millennia old Issue and Myth; as I described in my previous paragraph, if we think vertically and not only horizontally then yes, indeed, you pay no taxes in Arabian Emirates and everyone is also a millionaire in Norway, too . . . Adjust it in simulations by studying how and recommending it to those People, Government's and Corporation's to make the appropriate changes. Trying never hurts.

In my own Virtual Simulation's I have come across another interesting theory in Civ5 which is synonymous with this Future Scenario: 'Technically speaking if I, Silber, Psionic Warlock, regardless of whether we just blow 'em all the hell away by Domination, take over the whole Planet by 2050 we would A. Have no International

Trade And Commerce B. We would then definitely need other Planet's to conduct such Trade And Commerce with otherwise the whole Planet would go fait by the Not Existence of such. C. Where would different currency exchange values go to?

We are no longer slave to Nature. We are slaves in the work force, but we are no longer in fear of Nature in Modern Western Civilization! Even that god awful winter storm in North Ontario hitting my birthplace as we speak is relatively speaking to other regions and previous decades and centuries not killing anyone.

Now is the time to work with all the Forces Of Nature, not *against* them. There are other Forces besides Earth, Water, Air, Fire, Ether, Shadow, Light, Form, Spirit and GOD, like Electro-Magnetism which in case you still didn't figure out is what Null EM stands for. Soon, Electro-Magnetism and Gravity will be seen to be different manifestations of the same force. Then watch! Some Pessimist's say there will never be a Unified Field Theory and Theory Of Everything, and laugh out our work and development towards a European Union and there is no United States Of America because it's a Federation and not a Confederation, so once again they have the right to blow each other away, and in case you still didn't figure it out that is what the title of the 3rd Part of my Poetry Lore—1st Edition—Self-Consciousness, Law Of Unification (all bodies)—Published by AuthorHouse UK.

'Only when you have a far greater Enemy will you ever unite.' This expression, however, is strangely enough not working to well in the cases of Not Unnatural Disaster's, except for the great Community Effort and donations where it does not fail across differences, since the destabilization of their infrastructure and defenses causes Oppurtunity Hit's.

Humankind, to use an antiquated term, you're even better off using the word Human Type, since we've now seen so much cruelty and injustice to date, is a phenomenally, incredibly resilient Creature. And an inventive, adaptive, ingenious, beautiful Creature.

We must make the best of each of our situations. People look up to others who succeed, who made the best of their situation. I, too, am a self-made man who worked himself up from relatively nothing in a basement in Toronto, Canada on blow suck Welfare.

You can also use the Future as your spiracle causal focal point where Timelines will once again diverge and/or converge. Or do they just go Kaboom each time? With the future as our cause rather than being stuck in the mire and mud of the past we can make it, we can develop it and we might just not only survive but tentuple our population.

Mankind's and Womankind's entire Evolution is displayed behind us. And, yes, it's like an Aries—Venus, or Pisces—Scorpio, relation most of the time but ss far as I am concerned, it is impossible, a complete 0 probability that Humanity will not succeed, survive and develop to more and better Tier's Of Existence, and all their tears, pains, blood, sweats, efforts, works, arguments, debates, battles, warfares and wars will have not been for nothing, as presented and proposed by these various points, evidences, facts, statistics and what I now like to call my Death Analysis Trend's = DAT EM, and then only the Future will prove it Post-Analysis Trend, thus PAT your DAT.

Anyone who believes the opposite and does not see these things since the dawn of the 21st Century which everyone is laughing out only with New World Order and Brave New World statements, most I actually do not disagree with myself except for some major holes in their premises, logic and arguments, is only primarily pessimistic.

This is the Information Technology Age.

However, next to many other breakthroughs, this is the Age of Self-Consciousness.

Aware

People have gained the Power of Effect, and this is how they live, continually reacting.

People who have guns use the threat of Death to put into existence their belief(s), their point-of-view(s), their ultimatum(s), so I am going to call this happily your Ultimatum In Return. I don't think we need to tolerate it, especially with Social Medias and Internet.

They wish to be the only cause and as a result, other causes have been lost, lots of them shot, razed and burned down to the ground like Library Of Alexandria, not that I'm Anti-Roman or Anti-Goth since these days we just don't resemble the Greek's and Roman's anymore; they also had no Internet which is always a cute little retort to such texts, I mean symbolically describing 'interconnected web' is still a Native Indian statement and we don't even see that one elsewhere with even the words we use today like Computer, Intranet, Internet and very many other ones.

It has always been the people with guns who have been able to determine Existence and Evolution with their Power to decide who gets published, prosecuted, imprisoned, hung, burned, drawn and quartered, dismembered and/or decapitated or not . . .

It is categorically wrong to force others to follow one particular way with the threat of Kill, Murder and/or Assassinate and/or any other Form and/or Fashion of not unidly threatening anyone, and not just on Internet these days which seems to be totally in since everyone loves to Bash each other all the time from Hollywood to Bollywood and Amasia Entertainment and back again.

However, you must never forget that a 3D Game on Internet, like my Generic Universal Roleplaying Game System (GURPGS) with such Member's, Character Classes, 2D/3D Object's and Values are nothing but such; if you're one of those Nüber Püber these days who is Halluci-Nation all the time and cannot disceren between Reality and Fantasy anymore, like everyone else in Hal Mode, then get a Chill Pill or something.

Everyone will obviously choose Life over Death and just bullshit, resist, rebel, revolt and do you right back in via their associates over time: This is across the various first Family Values going all the way to the first apes with clubs and throwing dung at each other actually the very not unreal cause of all your conflicts, 'I no yo did my cousin in in 1713 en dats where I duh the line . . . ' It's not even Maffia these days, it's plain old Family Protection Values since he/she knows you did that across spying going back millennia but has no way to prove it with only Not Admissable Evidences, Killed Witnesses and plenty of Technicalities which tosses their cases out on their asses.

And so Killer's, Murderer's, Assassin's and my own favorite Spy Killer's, those of evil intent, have been able to determine the way since the very first apes even. To back this up have a look at (and in if possible) at their Ape Genealogy Timelines . . .

Now, let me ask you a question: Is it Life to do what you want? Is a threadbare of an existence Life? Would you cause Death to get the Hell out of your miserable existence?

Any of those Individual's, not that one can define this Object since each and everyone is already different by the Law that Two Identical Object's cannot exist in the same Space Time Continuum, who wish to live by the way of Ahimsa, the way of non-violence, go together and oust from Power those who would deny others self-expression; I'm not per se for or against this approach of Peaceful Protest but often it does not fail eventually.

Do not adopt their ways of Death. Then you become them and lose. The error made so many times, especially in recent years and decades, is that you the Violent Protestor's actually give them a shit load of Ammo, Shield's and Live Ammo (thus Live Broadcast Ammo) to stand on their heels and call upon every first and last Rule, Regulation and/or Law of their Government through Amendment's, Parliament, Right Of Ascension, Revolution, Official Document's, Debates, Battles, War's, History Of Humanity, Property, Ownership, Intellectual Property, Copyright, Registered, Trademark and this one I like best of all, Bloodline, which never forget Europe does not lack through Nobility's.

Go and using every Energy, gain the Power of numbers, the eager surgence which will break the tyrany of imprisonment.

Go to your Government, who is supposed to be the Representative of the Will of the People, go to the Corporation's, now listen to this, go to their front doors, windows, walls, their Head Offices with the numbers you have collected, and talk to them. Does anyone want another century of deaths there in Ukraine, either? Often if you even have enough numbers they will not fail to get around the Knight's Round Table and talk. Even if you cannot reach an agreement through various middle grounds and/or concessions in even a month or a year it is still better than all the useless senseless bloodshed like you're a bunch of Blood Cannibal's and/or really do go back to Planet Of The Apes Complexes which have not failed to dominate entire History Of Humanity but we are now in 21st Century with the key difference of Information Technology, Modern and Quantum Science and Technology and we would like to ask you if in a lot of cases you really want to call yourselves 'Modern' and 'Civilized' when it's only 'Modern Warfare' . . . I mean as a bad joke which I cannot resist repeating myself in our 3D Chat Environment called Noobie Chat Mode and/or No Noobie Chat Mode do we each just suffer the worst Compulsive Obsessive Disorder's (COD) with the stamina bar going up and down like a crazy motherfucker and woohoo 20:5 Kill:Death Ratios but it was still a lot of fun . . . And who would go and wreck a perfectly fine acronym, for all your own DOC's too, whyever is it CDO, again?

If after knocking on their doors, they show they will never change, are totally adamant, which must be seen over time, if they do not listen to reason, to the Evidences, Fact's, Statistic's and Truth, and you have done your best, your Group's, then do not shed their blood or smash their windows or even try and bring down their walls.

Do not harm a single of their souls. If you harm a soul, something living, then you will have adopted their way of Death, you will have become it and you will fail for you have now granted them all the rights to stand on their heels and call upon such things just mentioned.

Do not break their windows, doors and/or walls down for that is Violent Protest with all of its consequences and is not Peaceful Protest with all of its delays. Still though, you as an Individual, Group and/or People do not lack the Right Of Peaceful Protest. Even in their own systems, except the most extreme Totalitarianism's, you do not lack this right.

Now, be warned, again, if you harm a single living being, either by insulting them, battering them, shedding their blood, killing, murdering, assassinating them and you lose the purpose of talking to your Government or Corporate Representatives then this whole world will be lost. If you don't believe this, then you are dead already, since you unfortunately did not get this document, as of yet, or blatantly ignored it by their or your own Evil's Of Ignorances Effect's, as so many are repressed and suppressed to date, we do not lack that problem here getting through the somewhat not-so-innocent delay in Film's, 3D Games and other Multimedias to date, not to mention all of their Lies and Rumor's and Lies And Impersonation's and lame blatant attempts to Image And Label me, like other Celebrity's, which I vehemently object to each and every fucking day .com, and no, I don't need to swallow or get used to any of your Insult's And Provocation's each and every fuckin' day .com in your Medias from those on the complete other side and half of Hollywood; there is no other Medium but Internet for me and honestly why don't you go and blow suck hurl your whole outdate crap Analog Radio System's for once since that is pas god awful Cross Signal Interference.

Do we need to tolerate their continual bashing and ultra-lamin' blamin' shamin' attempts as a bunch of 12-14 year olds since beginning of Hyvescool? If that is all you do under the guise of Free Speech with all your Personal Attack's, never even arguing anything, then why don't you go and target your Target Group cause that's all you are and I do not give a flying hoot what you think of me, what your Unofficial Opinion, Official Opinion and/or Opinion is of me and/or anyone else; my self-image is also not based in that of the Enemy; the daily Crap And Ignorance coming out of your mouths continuously does also not offend me since it's not based in the Fact's and/or Truth.

Your purpose is the purpose of Life. This is like the breaking of the Berlin Wall or any other uprising in a long line of Fight For Freedom's through entire History Of Humanity.

Go to them and say or leave this Note:

'You are denying others the right to creative self-expression. Your ways (the Government and Corporation's) are destroying the world. It has now become the will of the people to preserve the world: Look at the consumption of explosion of over-population across total outdate infrastructures and to repeat from my previous chapter: Think vertically and not only horizontally across Appartment's and Animal Appartment's.

If this mandate was not broken then we have not broken down your windows, your doors, your walls, shed your blood, killed, murdered and/or assassinated you.

If you have not granted due process and procedure and audience and Debate And Discussion to the People then you have not followed the mandate of the People, the mandate of Life, of the inherent right for all Sentient Creatures, of Low and High Relative IQ Level's which is not lacking in being defined i.e. your Specialty Level, to creative self-expression, to co-exist peacefully next to each other, rather than Kill, Murder and/or Assassinate which is primarily Battle, Warfare and War. You will follow these priorities which will preserve this world which are the will of the People.

Since you have the Power and/or Energy and since for decades we have given you our Money, our Votes and/or our Follower's, you must, however you can, do the following:

1. Self-Sustainability for all of our and your Children must be Priority 01.
2. Primarily non-polluting and renewable sources of Energy must be eventually used as a slow development out of the Fossil Fuel Age to the Hyper Modern Energy Age with an approximate estimation of 50% of entire Planet Earth by 2050 as according to such Energy Contract's signed.
3. Animal's must no longer be tortured as shown in many Medias. To satisfy our palattes, the better alternatives must be given immediately: Animal Appartment's, I may remind you, are 32 by 32 storey buildings with Hyper Modern Science And Technology applied i.e. their pens with that size of a building can easily be 2-4 x the size and throw in even a track around the pens so they can walk for some exercise. See Planes Of Existence—1st Edition—Published by AuthorHouse UK for the rest of the description and argument; this is also not failed by the support for the decentralization trend in the Near-Future. See previous chapter.
4. All Tool's of Death And Destruction, except those needed for Self-Defence and Construction, all such weapons as stated by very real documents and conventions, must immediatelly, across the world, be destroyed, be made useless, turned off, melted down, reduced to scrap metal and/or recycled. There is no productive purpose for these Type Of Weapon's except for self-destruction and/or obliterate the neighbors, again. This one is supported by many Government's and Corporation's in 21st Century and not only People, one could almost call a Vast Majority Rule on it . . .

6. Following the elimination of all of such Arsenal's Of Weapon's, which also cost this much per hour to Government's, Corporation's and People, almost every Economy at a min of a Major Victory Rule, thus 66%—75%+ of everyone on Planet Earth, which does not fail to Veto their Veto Rule in United Nation's, this is a Debate Issue which is still unresolved by them as we see another 51:49% Corporate Effect again, an unacceptable Minor Victory, will unite together in the common cause of repairing all damage done to Planet Earth, try to recover from National Debt's and Deficit's without spiralling out-of-control into WW 04, The Gulf War leading to 911 was already WW 03, and setting up systems for the future prevention of damage to Planet Earth and People living in its globe, especially with self-sustainable growth Economy's i.e. it has often been argued that we would be better off with a slow, sure, steady gain system and not Bust Boom.

7. Following these, all People should be allowed to do whatever they want as long as they harm no one.

8. Free Market, Free Speech, Free Expression, Free Enterprise and Free Democracy are the rights of all Government's, Corporation's and Peoples. See a comprehensive system of Free Democracy in a later chapter, otherwise technically speaking Modern Democracy in Modern Western Civilization does not lack already the vast majority of each and every mandate here.

Unfortunately, as we each see, Rebel Government's, Rebel Corporation's and/or Rebel Individual's, like all Human's don't have a Rebel Spirit from a Child already, persist in killing, murdering, assassinating and/or destroying everyone around them. I still simply argue that you're also effectively doing such to yourself as a Parent and your Children.

If you, a Government or Corporation or People, do not immediately follow through with the above absolutes, make the first step in the direction of correct self-sustainable development and preservation of Planet Earth leading to actual growth and you deny us now then our numbers will inevitably grow, for this is the will of the People, if you don't realize it already then your entire Civilization will be at risk and eventually be brought to a standstill, and then even go in reverse by loss of Economy and equivalent Retrograde Effect's back to the Feudal Ages, like they do badly suffer from in 21st Century, effectively resulting in your total loss of Power And Energy, not because we per se interfered and/or tried to intervene on your behalf even, but by definition. You will be unable to prevent this by your not unidle Threat's and/or Act's of Death because this is the will of the People, to live in happiness, of each Individual in a Group. It is, in fact, inevitable for it is impossible by definition and numbers to stop 100's of millions of people and before you know there will be many more billions who all need and want the same things.

There is no Martyr, Hero, Leader, Rolemodel, Figurehead, Individual, Celebrity, Group, except those Group's of Government's, Corporation's and People, who spearhead this movement with its mandates, believe it or not. So, just because I wrote and defined this Ultimatum In Return by the will of the People through even my entire lifetime of 40 years now with a lot of Research And Development with the participation of many others in our 3D Chat Environment's on Internet, Internet Radio and even some Radio in The Free Show, my excuses if I now throw Near-Infinite Variables at each of you does not invalidate this document out of some kind of Only Subjective Reality Effect.

The above mandates are the mandate of the People to live in Freedom, Peace and Happiness on Planet Earth through primarily self-sustainable development with also the rights to colonize Colony Planet's.

So it is. So it will be. So be it.'

This by our many trend analyses and simulations and simply seeing all the World Event's on International New's Network's and many other Medias and Multimedias through multiple Medium's will work. It can be started on right away. It can be done without threat to the Individual. Initial impact is not important for with the dawn of Internet it already hit the worst Hydra Complexes. The Cause And Effect's are the deciding and determining factors as to see which Timeline(s) will dominate. Word will spread and as predicted the numbers will inevitably grow because the people are 'Fed' up whith nothing being done and primarily degradation, devolution and destruction seen since the worst Budget and Economic Crisis with raging War's since Empire State Building and WW 02.

The only Constant is Change.

I will end this chapter with this statement: 'Non-Violence is the most powerful and energetic force of change; more Violence only causes more conflict. If used for a creative purpose, even in Peaceful Protest, without harm to a living being, then it is beautiful, is not harmful, is Life and will create not just Revolution but Evolution to the Future of a Human Species who cannot fail to even tentuple their numbers and populate the Solar System and eventually the Galaxy. Only GOD knows when we will populate other Galaxy's . . .

Would you rather be in Death where both sides just keep killing each other?

Meta-Science

Introduced with One With Nature is Meta-Science. I would like Meta-Science to be the uniting of all fields of study to the promotion of Human development through Evolution. The acronym is also a purposeful propositional theoretical prototype as part of this whole thesis of mine to round off my York University: Do we own Nature? Better yet, which is less abstract, do we own the Solar System and since there are no Alien's in site do we own Milky Way Galaxy which gives us a completely different perspective then only Reseller Galaxy which states it will not fail to be very compy campy and Spliffed Down The Middle with the necessity of a 3rd Party Rebel Group. No, I mean, once again, as a Human Species with to date no Alien's showing up, do we not own the whole Galaxy across specieism and millenialism, not at all racism and centurionism.

The word Meta-Science is all about Human development through Evolution. The prefix, 'Meta', meaning 'along with, after, between, among' refers to Matter's and Energy with the promotion of Human development through Evolution. The adjunct, 'Science', meaning 'to know, observe, study, experiment, systematize' refers to our progression of Knowledge. Meta-Science does not fail to use Observation and Experience. Meta is better known and associated with the entire Aristotles to Plato Topic's and Debates and Pre-Socratic's does not lack in being my Specialty in my 2nd year of Philosophy at York University next to the 1st year of Psychology. I say also round it off, not specifying the BA, MA and/or PHD Equivalence since A. I, myself, cannot decide that or mark this and B. The now total ripoff costs of going to a University these days prohibits me from going to one for the rest of my life, I am now 40 years old, not only by my wallet but by principle; are we seeing another one of those Classic Lines Of Power Effect's where only the 01% Rich Elite get to determine all of Reality again . . .

Meta-Science is the study of Matter's and Energy's to the realization of Human development through Evolution.

Nature is composed of Matter's and/or Energy's, some say there is only a continuum of EM Field's but I still find it a little hard to swallow no Particle And Wave Theory, is it really only a fake sense of resistance and solidity through negative electrons or is there really a monad, and we are in Nature and/or in Reality itself which still does not lack despite all attempts to bullshit otherwise being One Big Reality. By definition we cannot be external of this Object as a Subject. Also a very common mistake is to state we are 'on' Planet Earth, no we are 'in' Planet Earth, an Object which does not lack a 4D+ Hemisphere which we are in and not only on, I mean are we skimming off its surface like throwing pebbles across a pond . . .

Meta-Science is the study of becoming One With Nature.

The system, the tool and the method, of Meta-Science is for the development of Human through Evolution. Its Law's are from Nature already. Many have also stated that the terminology 'man-made' which should even read as 'human-made' is completely bogus since Nature did not fail in making Human through development in Evolution already. GOD is still a Moot Point and is no not a Retort; there can never fail to be A = A Point Of Origin Of Creation and that continuously Time Loops on itself.

It is very important to spend as much time outside as possible, this way your exposure to Nature is vastly increased, not to mention get good dosages of Vitamin D and fresh air, thus obviously not hang out next to the Highway 401 while lighting up for too long, a little bit does not fail to help for Immunity's. This is where Meta-Science must also be conducted, not just in Private Laboratory's and/or Offices and/or Homes. It's actually not too over-complicated in principle though it does not lack Near-Infinite Variables in practice since you can also badly error if you have only double double controlled experiments, then you keep asking yourself why it keeps failing in the field and try to keep blaming it on Technical and/or Human Error when your premises are unsound. Remember: You must not lack both the Test Group and Control Group at a min of.

It is incredible how much time we spend inside, some do never go out anymore in whole huge Luxury Appartment Complexes. To date in entire History Of Humanity, with also total blow insulations, practically everyone was outside every day. We now have new causes, symptoms and syndromes from 20th Century, my Top 10 Favorite Classic Science Fiction Techno-Thriller still being Count Zero, which we have no clue what the short term and/or long term effects are or will be . . . In these cases, there is NO Control Group's, in fact it does not fail to suffer from the worst excessive quantity and quality of totally random Test Group's which results in nothing but Hydra Complexes.

The quality of Earth, Water, Air, Fire, Ether, Shadow, Light, Form, Spirit, GOD, as stated in my previous chapter, is far greater than you think, just take a step outside the door for the first time after a long cold winterr with your long pale hairy Scottish legs not having showerred, bathed and/or washed your kilt for the same perrrriod of time . . .

There are countless examples: We are getting in this Modern Western Civilization far too much filtered Air which is dry, we have much less contact with grass being part of Earth, many now due to fear of UV are getting far too little Sunlight even filtered through windows which us always at least at a min of 2nd degree reflection, and also now in some City's no more clean naturally occuring Water to drink and immerse in since it either has too many chemicals or too many bacterias. These example in 21st Century is almost already proof, not just evidences, and one does not even need to argue Ether across all of the Telecommunication's, Social Medias and Internet with the new phenomena which actually just happened to me about a week ago: My smartphone now has a faster Internet Connection and is faster than my PC! Holy Outdate Batman Effect!

There are sufficient parks though let's have more even for everyone to enjoy. Also a standard NL Joke I like which I heard in The Hague is, 'My god, where did that park bench come from . . .'

Nature Is Beautiful. Nature is Gorgeous. One With Nature.

We come from Nature. We are made by Nature. We are forced to follow the Law's Of Nature. Nature = Reality and it is all in One Big Reality. GOD = 1st Instance = One Big Reality, otherwise as many of you error you exclude GOD from Reality again. It is still a Moot Point since, once again, what are all the other Instances and what is GOD without his Henchmen? Thus, no, I am not in violations.

It is not just expounding the glorious wonders of Nature for all to discover, it does not lack a whole methodological system, and not ONLY Scientic Method again either, fundaments, foundations, means, manners, moods and Rules, Regulation's, Law's and Axiom's to be like an Investigative Officer, and be more of a Robin Hood instead of a Rebel Teenager, to find evidences, proof and test them. Unfortunately, as the Chaos Theory likes to have it, Nicht Ontheil Loki again, though I do prefer Thor, everything on paper never works in practice.

Do we need or want anything more than the next latest greatest shit since Einstein, the very real Math Formula to bump off at a max of Speed Of Light which I propose and present in Planes Of Existence—1st Edition—Published by AuthorHouse UK. I also propose and present a full comprehensive system in a later chapter here. Some say, I am still 50/50 at the moment, that Evolutionary Essay's is Meta-Science, a whole new system like that so-called New Math School to not fail at some point in the Future allowing for Time Of Development in discovering all the secrets of the Universe, not excluding very real Immortality of the Host Body which your Mind, Spirit and/or Sould temporarily occupy through Karma and Reincarnation, and not fail in a Unified Theory Of Everything which is how I like to call it rather than their not unofficial term and title for it, and I hope I'm not ripping off anyone by accident, credit due where credit is due, and if oopsy I get blamed for all of that shit again, well don't forget to give me all the credit too . . . In case you did not read it already, Evolutionary Essays will not fail to cross-reference as much as it can while you each also have Department Disease, Speed Of Shadow is near-instantaneous and does not suffer from the limitations of Speed Of Light since such being Holes and Rift's there is also no resistance and/or mass factors: You just go there.

Well, as they say, once again, always leave something to the Expert's, I've always sucked in Mathematic's since Hyvescool for no other reason that I'm simply not a Beta Type, not cause I'm stupid or lack my Science, Technology and Information Technology, I even did half a year in all of the Beta Specialty's up to even last year of Hyvescool with extra night course and also in Geometry, Calculus, Statistic's, you name it, even in NL which is just no fun and not easy across US, CA, UK, IE, SCOT, FR, NL and DE again, I mean, OW, you're asking yourself where the cross-incompatibilities are coming from again, it's not even racists, neo-commi and/or neo-nazi in these instances, we just don't get along anymore again since we are simply Incompatible Group's. No, I just really am into, like, only Suck Math. I mean how bad can some Statistic's suck these days, ' . . . that's not a couple people, Sir, that's literally millions in these numbers in this many International New's Articles per day from the millennia alone . . . '

Just to poke some more fun, talking about another unpredictable Death Analysis Trend though many say it's 2-Bit Obvious, the Auto Traffic Congestion and resulting symptoms and syndromes of staying *too long filas* in Fossil Fuel Age are also off the scale.

Therefore, we are also only forced by definition and such existing evidences though nothing is, as of yet, really constituting proof of any kind for we do not know by definition what will happen in the Future with these Death Analysis Trend's until we can Post-Analysis Trend's, don't forget to PAT your DAT's, to primarily rely on above mentioned methodologies and use primarily Reduction System's. This is an official School And System in not just Philosophy but also Science, Technology, Mathematic's, Information Technology and many others, I could almost argue if you keep trying to repeat argue Moot Point's, 1st Instances and try to ignore, sidestep, circumvent, preclude and/or a priori it all the time through denial of Rationality, that you actually do not lack it, tell me that your 1st Instance ONLY position is not actually using reduction all the time . . . I find it cute these days, all of yours and ours Basic Self-Contradiction's and I try to as often as possible quote their very own paragraphs against them: Is not Original Error not equivalated with Original Sin already? We also have our factions fractions fractals fractures fucture factals displays, you know, this part of our anything is not per se that part of our anything.

Besides parks in City's, a superb option is nearby preserved Areas around the City's. Go there by foot, bike, tram, bus, car, train, boat and/or plane even, I mean talking about missing the Tourist Potential again . . . it does not mean wrecking pavement, parking and/or building by the growing roots and branches of Trees, how many times have we heard that lame attempt at a retort, you can also as I stated in Spy Kill's 02 selective chocker block with a border of Tree Lines like UK Crop Field's the Amazon rather than raze the whole thing down to the ground cross your fingers, hope for the best, Santé Maria, give a little Prayer to your God, knock on wood and wish that when Monsoon Season and/or Hurrican Season comes again that it doesn't all get washed away and not only completely wreck NA in return again instead of only dissin' SA . . . This to us is 2-Bit Obvious Camping Politic's And Economic's Bashing of the worst kind since which Colony Century even when the first Republican's, Confederates, and Democrat's, Federates, decided to turn each other into the Numero Uno Worst Blood Cannibal Enemy Nemesis and conduct Civil War's on each other to date bringing everyone else into the Freya's and now pulling entire Modern Western Civilization down into the worst Lucid Nightmare Deficit Complex since the Stock Crashed with Empire State Building free style base jumping and free falling that if the Debt Ceiling defaults, how many times can it be raised in the eyes of their own Loan Shark Bank's, and the US Dollar is no longer the Base International Currency switching to the European Union Money System across EER, EC, EU, Commonwealth, UN and

UK Issues it will not only inflate entire Planet Earth leaving only us here in UK and North EU Region's with a single red cent to throw anything at but it will Lag Effect the entire board so bad you'll think you're on Totalled Outdate XP Homey Boy Hardware And Software again . . . In fact, it could also not fail to actually trigger WW04, I've been saying it for years now even, they will go looking for their Money . . .

Each keep forgetting so quickly the crucial factor of the Happiness Level of your Civilization, which if you ask me having done countless Civ Simulation's since Civ2 is the Numero Uno Win Factor to date, which do also not fail in generating Money and stimulating the Economy; I don't know how many times I have to says it these days in our 3D Chat Environment and/or Internet Radio alone: Each coin has two sides. Spinning on its edge is doomed to fall at some point in Bust Boom System.

Thus, what is wrong with the 3rd Party Liberal Government and/or Corporation? This is definitely rhetorical here since that would require a very lengthy response.

Think about it, instead of sitting in front of your computer all the time, like I do to much with resulting Every Other Day Bad RSI Symptom's, should you not go out more, one can the take in the direct warmth, Light of Sun, breathe in blue fresh Air, touch ground, grass, Earth, with bare feet, and maybe even immerse oneself in not chlorinated clean Water, probably now only in 2014 CE lime based lakes, instead of being hooked continuously to My Free Show = The Free Show, cause if you miss a minute you keep wondering and worrying if you missed something, with this many Every Fuckin' Day .Com Event's across Privacy Public Issues along it's no exaggeration, and your Ether, definitely duck tape her head, poke one suck hole and screw a smartphone in her skull, one can then admire the expanse of pure blue skies with white fluffy clouds rolling overhead around Mother Earth, even in winter it's a breathtaking sight . . .

Gaze at the majestic Trees, watch the swirling gurgling Water, witness the interaction of Plant's and Animal's and Human's, relax, socialize, and have for once an enjoyable experience, instead of all the daily let's Bash Thrash And Trash everyone and everything all the time, that doesn't even help, with all the infighting going on, it's no wonder that the East is taking it all over again: Will we even maintain and ever regain ownership over our own Canada?

One can enervate our Life, having regular doses of pure Pranic Energy induced into one's being. We also don't need and/or want our systems, businesses, Government's and Corporation's ripped off by the East which we are now being threatened with due to the risk of faillisements in all sectors, except for a smaller percentage who think that we will get stinking rich by selling out to them, well, no, don't you think they'll just stick it in their own pockets . . .

Two primary arguments and/or evidences (and many other reasons) back this up and not just a little bit: A. We technically and theoretically have alread Unlimited Resources if we go about it correctly: Self-sustainability applied to all sectors cannot fail at some point in the future, except the consumption rate of America alone is off the scale. B. We also have Unlimited Reproduction: Think vertically not only horizontally, through Animal

Appartment's though not next to Granny's Appartment, again, cause it is a Major Sound Disturbance unless you somehow want to fork out this much extra moola for Hyper Modern Sound Insulation Layer's from Japan, and Plant Appartment's and all Types Of Production Appartment's, with inevitable decentralization, in the range of 32 x 32 to 64 x 64 floors and walls in width and height dependant on weight of building by the time we are done developing and constructing them near your City we will not even need the Space Colony and/or Mining Colony which still heavily Fossil Fuel dependant still takes a half century already to get back here with the first usable materials except for pieces of hermatite from Mars. I have one piece myself and it was relatively cheap to buy on Internet too which still one of my most prized possessions. Nicht Ontheil Aries!

So, convinced, yet? Open your windows and doors, though not your walls, ceilings and floors too much, and let the outside in, you will not fail to see a whole Plane Of Potential.

Truth

Yes, Truth does exist.
Those who deny it
Are mislead Conformist's.
There is Reality.
Are you denying the existence of such a Tree?
Then thou art crazy.
There is a difference between,
A Tree and a Dog,
Though they are closely connected,
If you know what I mean . . .
Yes, look at what I mean,
Average the connection.
What is true
To all things?
I will leave such up to you . . .
And be satisfied,
In revealing,
To you, both an Apple
And an Orange are Fruit.
True, is it not?
All things are Energy,
Truth, my associate,
Is Commonality.
You are well come.

TRUTH, Poetry Lore—1ˢᵗ Edition—Self-Consciousness, Law Of Unification (all bodies)—Published by AuthorHouse UK

This Truth is independent of the Overseer. It is the Observer, the Human, who is fractionated, who cannot see all the Truth; he and/or she has only and always their own sunglasses on in Subject Reality and never enters Objective Reality which technically speaking only a God can see. As to whether this is Time bound is still highly debatable though is most likely since you would then be Immortal, no longer limited by the Cycle of Conflict, Suffering, Death and Life which is also primarily caused by your confusions in seeing through only your own glasses . . . That's also why I have 5 3D Film Glasses next to my computer on top of my Canadian Flag, you'd think they'd have figured that one out instead of all their finger pointing.

CAUSE AND EFFECT, Poetry Lore—1ˢᵗ Edition—Self-Consciousness, Law Of Unification (all bodies)—Published by AuthorHouse UK

'The trick lies, in discovering, you fool the trick. So grasp at the underlying, and you shall see Light.'

'Truth can be seen as Light: Light is between all Matter's . . . The way to discover Truth is through Observation and Experience. Observation and Experience is what all fields of study have in common. Some have only Observation, some have only Experience and many have both as their method to discover Truth.

This is the problem of all studies. This is why everyone is bickering over who is correct, who has the correct method for going about our ancient, timeless, genetic Quest for Truth, a Holy Grail Quest equivalent with the Quest for Immortality.

Well, Truth is Now Here.

This is why I recommend an uniting of all fields of study into one: Meta-Science.

"How proven?" you ask.

"Well, this is The Question, is it not?"

(A pre-Socratian-like discourse)

With people attempting to manipulate others, with the many levels of Nature, and with people's desires to deny Truth, Truth as a guiding proinciple has been lost, become muddled, and been given up on in Despair.

And Hope with it.

People have come to say, "Truth does not exist.", "Truth is only relative.", "My Truth is different than your Truth.", "Truth is useless." And people, worst of all have come to say,

"Truth cannot be determined or shown." and "The Truth hurts." And even more terrible, "He who speaks of Truth is a liar, looking only for self-gain."

I also like this one which is actually quite true: "The best lie is a Half-Truth."

Common Statement's in the Modern Age

Something which is in common with something else, which therefore has commonality does not have the same as something which is common and/or a commoner though they obviously do not lack Truth = Commonality with each other.

This is treacherous pathway . . .

With Truth denied by the people, no solution can be reached for any problem. And there are lots of problems.

"How have I proven?"

"The only way Truth can be proven is by not contradicting the nature of Truth itself."

"What is Truth? What is True?"

"Truth is Commonality. Both an Apple and an Orange are Fruit . . ."

(A pre-Socratian-like discourse)

Likewise, all Trees grow leaves, all squares have four sides, the Sun is warm, and so on.

Truth is the Common Quality of any thing, though definitely not Common Quantity unless you equivalate Truth with only numbers which to us is no more than an empirical fact.

'Common' means 'belonging equally to, or shared by, every one or all.' Thus, 'Truth is Commonality.'

By the fact, I, little me, discovered this by myself (and the words of the some wise sages) shows Truth can be discovered by a Human being.

And we are little, tiny in fact, 'Go to the top of the Universe, look back down, there Planet Earth is all gone, woohoo, problemo solved . . . '

'You are nothing but a nanoscopic particle on an Infinite Timeline.'

'You are nanoscopic and Planet Earth is microscopic, look at all the thousands of Galaxy's, however, what is wrong with this picture, despite the tiny size of your Host Body and only a few have made it up to microscopic Global Level, was it not always to date the Ideas and Ideaology's which are so much bigger that decided all Timelines . . . '

There is another Truth:

'We Are Capable of Comprehending All,
Of the Universe, Everything,
And Relating It to other people.'

ALL THINGS, Poetry Lore—1st Edition—The Power Of Release—Published by AuthorHouse UK

So, whether you like it or not, Truth exists, not because we say it or they have decreed in their Infinite Wisdom and Near-Infinite Power And Energy, more like only Lines Of Power to do a little bit more than just poke fun now and zen, but by definition.

Nicht Ontheil Veritas! Nicht Ontheil Mercury!

However, because the nature of Truth is not contradictory, you benefit from Truth. You do not benefit yourself by contradicting yourself, unless it is for a deeper Truth, for there are many mirages and barriers on the path to absolute Truth.

After all, Satan, Lucifer and other Evil God's and Goddesses like Loki and Luna, depending though a little bit what side you're on or what phase they're in, do their best to mislead Humanity all the time, after all did not GOD give you each the best and worst gift of them, the capacity to choose between Good, Neutral and/or Evil, Free Will . . .

Remember:

'No matter how much a Truth takes,
It will always give more Back.'

EMBETTERMENT, Poetry Lore—1st Edition—Self-Consciousness, Law Of Unification (all bodies)—Published by AuthorHouse UK

This literally means the Evolution of your Being, your Lifetimes, the growth of your spine upwards through even your very first Incarnation's as the East believe through Karma and Reincarnation and we do not lack in Celtic Baptism to support with Pantheon's such as Hinduism, Polytheism's and and other associations a theoretically not impossible Poly-Animalistic development of your Soul, Spirit, Mind through various Host Body's, if you interpret the Monotheism's slightly differently with their heavy emphasis on Aesthetic's, Ethic's and Moral's, with all of their own Do's And Don't's in Judaism, Christianity and Islam then you see it does not lack such either.

However, 'belief is not enough', the only way we will ever prove any Religion and/or Philosophy will be if can somehow, once again, show the the Soul or Spirit on a computer screen. Belief is fine for your Heart Principle and even Rationality, it's more like the world is at this time dominated by the worst Irrationality Effect's, our Modern Western Civilization since the 1st Renaissance is now so ingrained in Science And Technology using ONLY 113% empirical data again, that you have to ask yourself if it's not the worst not unbiased shit even. See previous chapter for a description on Invalid Methodology's about this Topic Of Debate.

This is the 1st Beauty of Truth: "With Truth you progress, without it you don't, to deeper states of less self-contradiction. Thereby, less conflicts, friction, pain."

This is why Truth is difficult. Going with Truth is the process of growing up.

'All things Grow Up.'

This acronym is also purposeful, it quite literally means now that we have various breakthroughs in Zero Point Field to be your Null EM Field's.

I have posed the Question elsewhere in my works, so I'll repeat it here, where are the Qaulity Values of you, me, each Living Being or Inanimate Object, in all of your Quantum Sciences? Well, where else would they be . . . They cannot be not in and/or around you and/or connected somewhere . . . How do you know they are not smaller than Nano Level?

In fact, is not Science and Philosophy not the revealing and/or pointing the the correct direction to make that which is presently Invisible to be Not-Invisible?

How do you know, therefore, that all such things still undiscovered and/or unexplained and/or not proven are simply smaller than Nano Level . . .

Then not even Science is in violation of Religion, and vice versa.

YOU HAVE TO FIGHT FOR FREEDOM, Poetry Lore—1st Edition—Self-Consciousness, Law Of Unification (all bodies)—Published by AuthorHouse UK

Once you grow up, though, once you gain a deeper level of Truth, and this is important, you never have to go through such conflict, friction and/or pain again. You can never lose, except material possessions, what you have gained, already.

'Now.
Now you can climb.
Climb, Climb
And never fall
On blocks of solid reason,
To the Tower in the everlasting sky.'

YOUR LAND, Poetry Lore—1st Edition—The Power Of Release—Published by AuthorHouse UK

This is the 2nd Beauty of Truth: "Truth gets you there."

If you choose not to solve your conflicts, frictions, and pains, you will keep those conflicts, frictions, pains.

And this is the 3rd Beauty of Truth, there are theoretically an infinite number of Beauty's of Truth, and also only one: "There is an end to conflict, friction, pain at some point in your Timeline, whether that be at point of Death or not."

'It can be the pleasure of growth from
Now on. Those who must suffer more will
Pass on. Our spines have become stronger,
Energy can flow easily along them, now . . .'

HELL IS THE DENIAL OF YOUR FLOW I, Poetry Lore—1st Edition—Self-Consciousness, Law Of Unification (all bodies)—Published by AuthorHouse UK

In case you are curious what the One Greatest Beauty of Truth probably is it's "True Beauty of Truth" though is a little bit a circulari. Like Narcist who could only love himself and stare to long at his own reflection in the water and be turned into stone until a Goddess came along to free him and/or her Truth can only satisfy Truth. People generate too many exceptions, ' . . . if you generate one exception then you might as well generate an exception for the whole thing . . . ' You CANNOT generate a Truth with a Lie, it generates only more Lies, just like fight Fire with Fire causes only more Fire and burns the whole building down again and if you try to fight Fire with its perceived opposite then it causes a shitload of steam . . . Justice System also attempts to expose Lies And Holes to reveal, backup and support evidences, proof and Truth. Once again, in the vast majority of cases, unless it's clear cut, only by Reduction does anyone know for sure, to mildly paraphrase: 'Anything reduced ad absurdum must be True.'

If, you will, it is Truth itself, coinciding upon itself, the Great One Truth, the Truth which encompasses all Truth's, which is Common to all Truth's, is Truth itself. After All, Truth is the One which all other things have in Common. Otherwise, you really are just stuck with Relative Truth's only like Relative IQ Level's for Specialty's: 'Does one need to know only one branch of the Tree or the whole Tree to gain Enlightenment and/or Immortality?' Once again, the Greek meaning of the words Omniscient and Omnipotent and other confusions between Latin, Greek, German, French and English do not mean the same whatsoever. So how do you know you haven't badly misinterpreted and/or translated again?

Technically speaking to argue it a little bit if the definition of Immortality A = A Timeless, thus not bound by a Finite or Near-Infinite Timeline which is Not Infinite, then God's and Goddesses, Angel's and Demon's or Demi-God Heroes have no problemo doing Time Loop's on everyone like a Dr. Who Series off the holler toller boller. And that one can definitely not be an 'and/or' where in the Land of Andor he a Prince who had to fight many Battles and War's to reclaim his rightful Throne once again became King and reigned for many a year for they are not at the same level in the Hierarchy.

Yet, by definition, this is not incorrect. Thus, Immortality is theoretically Not Impossible.

Nicht Ontheil Zeus!

The recurring problem again is as it is presently and apparently always except for GOD, God's and Goddesses and Heros and Villain's raised to godhood through their Epic Adventures, though how you Succeed and Fail to me is still of key importance, is that no one can bring back evidence and proof from the After Life Party.

Truth = Commonality, however if you cannot prove the Existence of the Object in a Class System with Common Quality's and Relative Quantity's then it is invalid.

And so what is needed is a System, a Study, which embraces Truth, not only number oriented empirical data. With an all-round embracing of Philosophy, Religion, Science, Technology and Information Technology, which unlike Literature, Languages and almost every other Sector and Specialty to date demands a Unique Object and Value System, we will gain Human Development. A = A. A will in our English Alphabet NEVER be the second letter of the alphabet B. B = B. ONLY when A = B and C = B will A = C. A ≠ B. I imagine except for the Quantum Computer BOTH Boolean Logic ' . . . like yes pls do stick a 2 into the binary system next to all the 0's and 1's and maybe it won't have to process such so much crunch crunch crunch . . . ' there is NO Alphabet who dare go and violate upon themselves so badly, except even in English the same word has multiple meanings and causes all kinds of confusions, miscommunications, misinterpretations and mistranslations which is the True Cause And Effect's of ALL Conflict's if placed in the wrong contextual object oriented programming language.

So, I recommend, once again, a new field of study, Meta-Science, encompassing all disciplines, if it is to embrace Truth, which Unites, which is Common to all fields of study. This is not in violation of Sector and Specialty System's as stated here in a previous paragraph.

Meta-Science has Observation and Experience as its all-embracing Methodology, who's One Law is Truth, and due to these quantities and qualities resulting in Near-Infinite and Infinite Object's, Values and Variables, it is by definition the study of becoming One With Nature.

Next to reaching Enlightenment and Immortality it has the potential of gaining Infinite Power And Energy which is self-evident through History Of Humanity what we are all in contest with each other for, next to the obvious needs and wants to Near-Infinite Defenses and/or Offenses, and simply just take their Territory's and/or Resources, which is inevitable by explosion of over-population.

It is too bad we just can't ONLY Debate rather than ONLY Battle, even better always a fraction in between, for all the killing, murdering, assassinating and/or raping, pillaging, destroying only wrecks everyone all the time, except for one instance the mythical never reached One Globe Planet Domination Scenario, would we really have no more International Trading Partner's then . . .

Unilateral Democracy

'The good will of Humanity, so much for Democracy . . .' One of the first Philosopher's I read was Jean-Jacques Rousseau who I paraphrase with this expression.

'Do not mistake the expression for the statement.' As to who stated this one you can also look up at Wikipedia or elsewhere on Internet; I have said many times if I need to directly reference each sentence even then it would add another 5500 pages.

This chapter is in response to a previous one, since the associations with Social Democracy have become irreparable; as I have also said many times despite their attempts to image and/or label me I am not per se on any side being no more than a middle middle middle player in all directions except the most extreme ones. We would like to argue since many now agree that Politic's and Ecomonic's and practically each and every Sector and Specialty should be approached with a per per Sector Module System for their faction is obviously by Education and/or Work Experience ten times better at it than we are even. For example in IT it is to date now clear that MS Client's are better than Linux and Linux Server's are better than MS. This can almost be applied to everything though they are conflicts and cross-overs due to competition.

Unilateral Democracy will utilize Head's of State, Head's of Government and Head's of Corporation's in a far more balanced and fair approach towards Democracy. It also does not lack Leader's and other Rank's in the whole system since also to date we do not see the dissolution of the Representative regardless if such is a weak link or not as I stated in a previous chapter. Once again, what is GOD without his Henchmen . . .

There can be no single Figurehead in Unilateral Democracy which is also major weal link and did not fail to trigger many War's in History Of Humanity even. Such an Individual can only be targeted. We are not saying to delete the President, Prime Minister, King and/or Queen as they want to do in NL even daring to Insult our very long Bloodline due to that Power Of Signature and/or other Right's to decide things in Social Democracy. If I may remind each of you that Netherlands and Canada plus many other Country's are still a Constitutional Monarchy as according to the MySQL 2008 database.

There needs to a reduction if not an elimination in the power of each Representative. The Representative is also always the weak links in Democracy. On the other hand, if that's all one has to do just like your worst 51-49% Corporate Complex is sign away millions of lives with his/her Power Of Signature than you're each no better than their Dictator who keeps laughing out the West for that very fact.

There are serious problems with Power Of Signature and Ownership. You are already the Boss, you are already God, 'Your Boss is already God so I wouldn't say 'no' if I were you cause you're 'not in the mood' if he/she requests you to do that . . . ' This single signature can even delete and/or kill millions of lives when we're supposed to stand for Freedom, Peace and the Fast Food way . . . , 'I own the World and, like the Pharoah's, you are all my slaves . . . ' One thing in this line which I just had to look up and cannot resist retorting to them is the Sun Salute is not the same as their Arian Salute.

Nicht Ontheil Ra! RA RA RA!

There needs to be more Power of the Peoples in the plural and not the singular. This has been an error in terminology to date allowing them to retort continuously 'Your People are not my People'. The present unfairness in Global Politic's is an atrocity. The present lack of Equality, Equal Oppurtunity (and Equalitarian Aquarian) across the borders is not acceptable. It's even worse than it used to be in 2014 when we're supposed to be since 70's in Canada and 60's in Star Trek and 50's in Buck Roger's to be not failing in Fight For Freedom and actual development of such systems. Well, despite the really great ISS, 'There are no Scan Devices in Universes.com' at ground level here still where they are badly needed in especially Medical Sciences. It's not only a question of Investment and Budget Crisis, how many times now have I also said 'You are missing your Millionaires' Market's so bad you don't even know your Target Group's', but there is no will to do so. If you do not have the Will of the 01% Rich Elite to buy it and sign it in at Legislative Level then the Will of the People is meaningless in many ways, not to mention your Grass Root's Effort's, except for a few Labor Union's. With this one I would like to also state to not give up on potential unification in futures.

'You have nothing to worry about from me, I'm not the Executorial Department . . . ' says Null EM Elven Silber, Psionic Warlock.

'I can ONLY dysfunction outside of my Hierarchy's but what is that kind of Hierarchy Bashing?' says Silber, Psionic Warlock, 5-Star General.

And then the famous one, 'What system functions without Immunity's . . . '

More protection of Employees is a very important mandate. These days you can get easily fired for a thousand different reasons with no personal protections, insurances and/or guarantees which is a plus side of Social Democracy since you at least have better Social Insurances, on the other hand though you cannot escape from it without winning the National Lottery due to Heavy Taxes, Cheap Labor's and Free Project's.

The increase in the Average Income (middle) is long overdue, while one man slaves away for piece-meal another is getting rich in his sleep . . . The Average Family which differs per Country or State, even though I could argue in Modern Western Civilization it is now amounting to no more than the 3-Person Family Unit with Pappie, Mommy and Little Johnny since Little Julie was sold on Internet already cause she screams hi-pitched decibel octaves reaching the 24th Degree Of Heaven Planes, the plus side is she gets to live in Norway, cannot even survive or develop on only Daddy's Income causing increased stress and over-work everywhere, not to mention the worst Credit Crisis, faillisements, poverty, starvation, homelessness and then . . .

More protection of the weak, sick, young and elderly is vital. Welfare and Disability Check's for these cases are so low, not rising with Inflation, dropping and even being deleted in some cases, in the case of Pension's and your age, you're better off being committed into some form of Hospital for the real Sicknesses And Diseases which such causes on a physical, emotional and mental level. Forget living alone, these Peoples cannot afford to do so, these days which is still an interesting difference between so-called Social Democracy's, Liberal Democracy's and Corporate Democracy's since here in NL you're not allowed to live with more than two and draw Welfare Check's whereas in Canada and America you can share with a dozen or so and all draw Welfare Check's.

However, when we have now recently finally received International New's Report's that a lot of people in New York have only a $180 per month Welfare Check than you are not even a Democracy which stands for the Right's, Freedom's and Liberalization of the Power of the Peoples which they also fought for since Independance Day's . . .

The destruction of extreme left, middle or right Totalitarianism should be a matter of fact since WWII, yet each days harvests a new wannabe Dictator. The promotion of Unilateral Democracy will also lead to embetterment for all, not excluding your 01% Rich Elite who will also Invest and/or Own in even clean renewable Sources Of Energy. As I argued elsewhere in my works the logistical load alone of Fossil Fuel's in Space Travel by weight alone will lead to a transition away from such since it is lagging the whole field and Planet even requiring a half century now to Mine Colony Planet Mars.

Elimination of outdated left, middle and/or right descriptives and the implementation of an Issue based Unilateral Democracy with a primarily Sector Module System is imperative for the 21st Century and the future of us all. Everyone keeps arguing loops with such outdated jargon, they all miss the Issues at hand i.e. practically no one in the West has a problem stepping over to the competition if they don't deliver my product in 24-48 hours with no damage or other black adders in the grass. People always end up voting on a hair-style or color of the poster and how attractive something looks rather than Issues of relevancy and importance in a Valid Priority System.

Better and more frequent voting, for 1 Vote In 4 Year's is *not* Democracy, on all Major Issues will better show the Will of the Peoples. Minor Issues are on a per per basis and case from Individual to Community to Village to City to Country to Continent to Planet Earth to Solar System to Galaxy to Universe.

One cute point as part of my Sense Of Humor Element's in my essays so it's not so dry, long and intellectual, again, is note how I purposely do not stick a 'The' in front of 'People' and I do stick a 'the' in front of 'Will of the Peoples'. This to me is precisely what Jean-Jacques Rousseau meant, not so much the allowance, lenience, tolerance and/or exception of GOD Principle which is still a Moot Point highly repeated forever.

In IT 2D/3D Object Definition Development Department delegations and Debates we also have the same Issues, it's like trying to argue .scot land extensions or .mer or .merc planetary extensions for does it reduce and/or change the meaning and/or definition again . . . I did not fail to utilize this myself after someone ripped off my previous Band Name, X-Machine, by simply registering .de instead of my .com and .eu when my Boss and me developed and launched The Open Markets, and again for Mein Own Unique Combination, The Open Market's, to indicate possession and ownership. This is far closer what The Free Show is really about, not your heavy emphasis especially in proximal zone to The Hague on Poly-Cracker's on Crack Cocaine Complexes. Now that we finally got International New's Report's about the Mayor of Ontario (get it) we can only crack up laughing since it is that new trend of Popularism.

Nicht Ontheil Mercury!

A Global Planet System (which unfortunately causes a conflict with GPS) of cooperative Politic's, rather than segmented Apartheid States and Country's ruined by greedy competition and power-hungry Totalitarian's will lead to sustainable development, rather than wealth pooling and Black Hole Effect's in 3rd World Country's and Modern Western Civilisatiation's.

Elimination of imbalanced purely Capitalistic Hierarchy's where 1 CEO or only 5 Board of Director's or only 13 Stockholder's can decide the fate of millions is required for a fair system of Democracy. Instead, a more horizontal geometrical Unilateral Democracy, where 1 Sector, including the actual Peoples in it, can Vote against another Sector would be better. Therefore, limitation of each Head of State, Government or Corporation in a double-check counter system is recommended i.e. if spraying crops is proven to be damaging to the topsoil and the Environment then the Head of Environmental Branches Of Blow The Planet Away Again can Vote on it with other relevant Sector's, if not all other Sector's since it could very well also be in their interests, on an Equal Vote System basis, pls not one more Weighted Voted System Joke again, therefure 1 Vote per instance of Planet, Continent, Country, City, Village, Community, Family and Individual, thus also for each and every Representative of United Nation's which is the governing body of Planet Earth as according to the Geneva Convention and its revisions to date. That so-called weighted Voting simply replaces the Dragon with a Hydra. Theoretically and technically speaking the only thing, next to obvious in-fighting, which could wreck this system is if very not unreal Alien's show up from Planet Xeebop again . . .

The most important mandate of Unilateral Democracy remains Majority Win's as in the precepts of Democracy and Right Of Voting which was also erroneously defined as 'Right To Vote' since I can't even get there still resulting in only 50-75% of the entire population of a Country who got their Vote in, my paragraph in The Free Show which defines the differences between Minor Victory, Major Victory and others in Debates, Battles and Votes states if you also only win by 38-37 and call it a suces in our recent

NL Akkoord, it's more like another Nothing Gets Done Effect, then should you not Call a Re-Vote and/or Re-Count just like Florida with Bush vs. Gore? Otherwise, your system, once again, descends into Spliffed Down The Middle, Wrench and/or Bottleneck Effect's or worse the opposite Totalitarian, Winner Takes It All and/or Top Heavy Effect's.

Deletion of expansionary War Effort's, as horrifically escalated from 1888 by the new German Keizer with the tragic death by Assassination of the previous Keizer who ruled for like only 6 months and almost finished and presented his own referendum as a spinoff of Napoleon, a direct result of explosion of over-population dating all the way back to God Emperor Alexander, The Great and earlier and directly leading to and causing WWI and WWII is always of utmost importance. We now see exactly the same happening in beginning of 21st Century, especially the last years.

Poetry ad Infinitum, Defens ad Absurdum : Defens ad Absurdum, Poetry ad Infinitum.

'As to whether I need or want to tie you each down with duck tape and poke a breathing hole in it so you can scream better to my Poetry is irrevelant for I, myself, and me also have no more than 1 Vote in the whole thing so I cannot be my own worst petty Warlock and/or Dictator.' says Null EM Elven Romulan, Silber, Psionic Warlock, 'I mean how 'bout each of you after a half bottle of your stuff only, do you have any waiting thinking time at all before you pull out your gun, my own Null EM Laser Shotgun leans against the wall next to me just out of arm range so I have at least one momentary lapse of reason . . .'

For only with concentration of our modern and advanced War Machine, especially in the 21st Century and beyond, on Home Defense will we ever progress . . . and at this time with all the Wiki Leaky Hole Effect's in each system we might as well forget Laser Military until half way through 21st Century with actual maneuvrable Laser Space Jet's which will even take decades to test various prototypes presently being heavily invested into and developed in the Space Travel and Space Science Industry's. Lest we risk a far greater evil in the N2 Bomb, a thousand times more powerful then any H2 Bomb, which I shall now define for each of you so you know the magnitude of the potential of not just Blow Up The Whole Planet Effect which they will first use it for but Near-Infinite Defense Potential Effect which it will not fail in in the Future. N2 Bomb stands for Null EM Bomb to the 10x-1000x magnitudinal Power And Energy of the Diameter Explosion and/or Implosion of the Null Point of any particle in return upon the heads of the Real Terrorist Enemy. If you think of it analogously as potentially equivalent to far greater than the Kinetic Energy of a pebble being dropped from Orbit and impacting a Noobie House than you might figure out your breakthrough in the Near-Future which will antiquate the Fossil Fuel Age. In case, again, you get stereotypical responses, well you have nothing to worry about since it will also trigger the Space Age and most of that is already Invested And Owned by (the) 01% Rich Elite.

The elimination of National Debt working towards a profit-making Country, like up to The Early 21ˢᵗ Century Of America even, sounds like a bad D—Documentary Title, is not a far-fetched fantasy. Many other Country's prove it is possible, like Arabian Emirates and Norway, if not these days the entire East, for the Government and/or Corporation to make a profit, even in the 21ˢᵗ Century, did you know that at the beginning of the 20ᵗʰ Century that America made a small profit of about $200 million, why do you think they were so reluctant for so many years to be pulled into WWII to save Europe and England, and now they are only the bad guys, labeled by almost everyone as a Global Police Army Force . . . Lest we disintegrate into stock crashes and warfare . . . Testimony of this is the +/—$8,5 Trillion National Debt of America in begin 2008.

I left myself a Near-Photographic Memory Bread Cookie Crumble Trail cause after a half decade only it was +/—$13,5 Trillion National Debt in late 2013 triggering the worst Economic Crisis in History Of USA since their stock crashes with Empire State Building. Internet has an Absolute Photographic Memory . . .

The reinstallation of Colonization on a purely Economic level is very necessary, 'The end of Colonization in the 20ᵗʰ Century was the biggest mistake in Modern History.' It's like they don't want to conduct anymore International Trade and/or Tourism do to their age old left, right, middle differences again. Nicht Ontheil Marco Polo!

In their so-called fight for Liberation And Independance only Corruption, Poverty, Disease, Chaos and Totalitarianism ensued, just like Geurrila Leader's taking over half of Africa again as we donate %75+ of Internation AIDS to them and vice versa.

For Example: Are we going to Colonize Mars? Or Moon? It's not a question of 'if' but only 'when' which now like the 800+ inhabitable Planet-Like Celestial Body's of which only about 18+ are barren rock Planet Earth-Like Celestial Body's which have been found in proximity to Planet Earth, meaning in proximity to the Solar System, is now leaning heavily on the Consy Side, instead of years and a decade only try more like 50-100 years before we find a single habitable Planet, land on it, survive, grow, mine it and bring Resources back to Planet Earth with Space Cargo Ship's.

Transition to a Global Economic System (does GES conflict with anything?) which provides more fairness to all Peoples regardless of Sex, Skin Color, Nationality, Religion, Philosophy and/or Right Of Choice is not a Point Of Debate (oopsy now I caused a conflict with POD): Without diversification of the Gene Pool we will not survive into the Future for unless they also want to be their own worst sense of an Über Mensch, I prefer to call like everyone Domme Kanker Apschen Menschen these days, you cannot send ONLY the best of the best, the cream of the crop, the Elite Scientist's, Artist's and Laborer's, Species and Races, you also need weaker strains for bio-diversity.

You also encounter next to Near-Infinite Variables, a Hyperbolic Vertical Learning Curve the greatly feared amongst Space Colony FANatical's the Chance Of Survival.

This makes overly Optimistic And Liberal estimations of only 2020 to be larikoek.

A Talent and Skill based system taking into consideration Education and Work Experience from the very first day of Hyvescool at 4 years old is the better alternative i.e. it is not allowed to offer different Salary Level's based on other criteria, can you really categorize and/or classify a Cleaning Lady who has to bend over everyday into the Noobie Corner, though not near my computer, as any more Lover Rank than some Stupid Platinum Blonde with 1 Year Com's Certificate at Data Entry Secretary . . . yet the differences here in most Country's with the pretense of Democracy is absurd.

Better providence of Medical Science, Medicine, Medication and Medical Supply through Country, Global and Unilateral Health Insurances, instead of primarily profit-oriented Corporations, like Pharmaceutical's and Bank's, is badly needed both at home and in other Country's, our Government's are being ripped off badly as the costs of Hellth Care everywhere in Modern Western Civilization are sky-rocketing with out-of-control Inflation Level's and very little actual cures resulting to date. Everyone's got something chronic, everyday you hear another store via via of someone with Sickness And Diseases, so I'm going to say it once again: What a SAD World.

Better providence of Welfares to those less fortunate i.e. a single Mother or a beginning IT Employee or anyone else starting a new Career gets Fired by a Boss or suffers personal problems and does not survive on $500 per month Welfare in Canada which last I checked since I was also on it off and on from 1992-1995 CE. 5-10 Credit Card's later and the Individual is homeless. You can look this one up to as a running cross-reference: Now there are +/—6,5 million Homeless and Jail Prisoner's in America alone. As a Bad Joke, how many are there in Canada cause it's already gotten up to +/-100 thousand in NL . . . Welfare in each Country has to allow for the Individual to pay for Basic Costs: Living Space, Food, Water, Heat, Electricity, Medicine, Education and Children! Post-Hi! School Education is still highly debatable, in this case I can only say I am of the opinion that if you also cause a Massive Post-Education Rift between Bonafida Gratis Hi! School and now you have to Pay For Everything then you're once again stuck in a Prism Of Your Own Design. Which ones are lacking in your Country? How else is Country supposed to build its Future to a self-sustainable system without collapsing back into Bust Boom Complexes which are now apparently spiralling out of control . . .

In addition, if the Individual is not given the chance for an Education, Re-Education, Re-Classification and Re-Integration then he/she cannot be expected to find Work even after incarceration, therefore leading to the Vicious Loop Cycle. If the Individual is weak, sick, child or elderly then he/she cannot also not be expected to Work without first Rehabilating, Regenerating, Rejuvenating and Resuscitating.

The Black Humor these days, in the case of many hopeless cases ranking in the millions who cannot find Work again, is you could also send all the Homeless and Prisoner's off to War as a 6,5 million Standing Army since they don't need training and would rout them all away to the other side of Afghanistan, but ok that's comin' across again as getting a Death Sentence for being caught with 1-2 grams of Dark Full Moon Weed.

We live in such Luxury And Comfort in the Modern Western Civilisation that we have lost perspective since WWII; it's hard enough with only pen, paper, TV and Radio, only. Such is early 20th Century standard and even up to the 80's for the vast majority of People of Planet Earth, not 'People on Planet Earth', and many Peoples of Country's, not 'Peoples in/on Country's', are still in (the) Dark or Middle Ages up to the 20th Century, now retrograded by War and Natural Disaster's back to the 40's, if not Roman Greek Times, with not even a single TV in Wipe Out Zones: You try sitting in 1 chair staring at a blank dirty wall ad continuum while bibbering and drooling with a Self-Medication White Straight Jacket where you can choose to press the Pain or Pleasure Button's yourself and keep screaming . . .

Such lack of Welfare can never be called Democracy, let alone Unilateral Democracy . . .

Major Decrease of Crime Rates, not 'Minor Decrease in Crime Rates' by decrease of motivation to commit Crimes is the only possible answer; the outdate 101 Classic Psychology of late 19th Century, early 20th Century and persisting to beginning of 21st Century everywhere by everyone of 'Reward and Punishment' is exceedingly antiquated, deprecated and degenerated for it is now resulting in nothing but the denial of our pleasure center in our brains through everyone's excessive Sensai of Law And Order: 'Their Liberation has no Liberalization and/or Liberosities.' When you're sitting with your last $1,00 and no more Credit Card's, it doesn't matter what the bloody Reward or Punishment is, ' . . . the particle will go looking for its particles . . . '

As part of the Major Decrease of Crime the quantity and quality of Fines, Misdeameanours, Crimes, Federal Offenses and Act's of Terrorism have to be decreased or your so-called Democratic State is also no better than Totalitarianism with over-flowing Mental Institutes and Prisons everywhere. Nuthin' like Cheap Labor buth that's ridiculous! Even more absurd is ONLY Rule Of The Gun.

Primarily in Drug's Are The World 24/7 Cornerstores and massive Pharmaceutical's the sentence for Sedation Nation and Pleasure Middle should be a lot less. If Everything In The World Is A Chemical is your Country not leading under Double Hypocrisy? Am I a Criminal cause some bored Cop caught me with 1-2 grams of fully dynamically grown by the Fire Full Moon Marijuana? What about Marijuana for Sex and/or Medication? How about even Cocaine for Sex and/or Medication? None of the above blacklisted substances were illegal until the beginning of the 20th Century. Even cola had until then a small amount of coca in it. Banning such substances does not stop Peoples from taking them, especially in Europe and America. Rather, it makes such a Crime in Black Market's with no control over quantity and quality of crop, ingredients, production, delivery and/or sales vendor, and your Country does not earn any Tax off of them.

I have also often thought to date: 'The only Twoo Therapy I need is a good Blow Job so you suck the Poison out of me . . .'

For those in Jail, already, all Minor Offenses should be deleted, Major Offenses reconsidered as to the actual Degree Of Severity Of Crime of such and Education, Re-Education, Re-Classification and Re-Integration take place. If there is no Living Space, Food, Water, Heat, Electricity, Medicine, Education and Children provided then what does your Country or Justice System expect if they end up right back in the same Vicious Cycle, 'Here, let me load the gun for you . . .'

'If you eliminate the Cause then you also eliminate the Effect . . .'

Better providence of Safety and Security is lacking. The Police System is supposed to help you with this like Neighbor Conflict's but it tends to worsen the situation even more because then they Suspect me instead like a bunch of Wannabe FBI Agent's.

Instead of giving out Fines all the time, the Police System, I have now also been wrongfully hit with insufficient cause and no witnesses except themselves scratching each other's back so much they're bleeding out all over the place since Sep 2013 to Feb 2014 for my so-called Beer, Wine and Alchohol Habit And Tradition, all Lies And Rumor's fed by the NL Media and Poeparazzi with their political puppets, now also even Neighbor's Of A Celebrity, for €432 . . . !! This is so 2-Bit Obvious that they set you up with their hopes at the same time that they will finally Kill Your Wallet and Wiiiiiin The Issue, should be seen as the once was were and really *not* your 'friendly neighbourhood Spiderman'. The original purpose of the Police System was to protect your Home, Family and Friend's, to Protect And Serve not Corrupt And Enslave. Do we really need another blatantly innocent African in Jail or another wrongfully suspect Muslim Terrorist? Do we need to suck more blood from the Peoples since Colony War's? Did we still learn nothing from Washington, Locke, Jefferson, Martin Luther King, Kennedy's, Malcolm X, Clinton, Obama and all those who struggle for Freedom, Peace and Liberation . . .

Do you deserve to get shot, locked up, fined and given a Life Long Criminal Record for running with another Traffic Violation? DD'd, Dod'd, DOA'd and DOA?! Is not the Degree Of Severity a little harsh? To cover their own worst biasses again with long standing fights between left, right, black, white, poor, rich, hippies and hicks they now have the worst basic self-contradictions all trying to blame each other for what is no more than another Modern Symptom called Traffic Congestion, with obvious examples like if you have that Soft Drug, Hard Drug, Beer, Wine and/or Alchohol Blood Count in your veins and in your breath, go and hit your own Target Group and go and get the real Criminal's on Internet where they all are, unfortunately Police System is not Cyber Crime Department to date 05-03-2014, where and when everyone is going in the same direction at the same time in another suck hurl blow 9-5 Outdate Mentality . . . Well, you're just asking for it, that already ups the odds to accidentally impact other Traffic by a stupefying degree that I'll call it one of my Top 10 Worst Suck Math.

Staggered opening and working hours from i.e. 05:00, 06:00, 07:00, 08:00, 09:00, 10:00, 11:00 and 12:00 with a 6-8 hour work day and/or night does fail to practically solve all Unemployment even since it also provides All Types Of People to work Part-Time and/or Full Time depending on their Life Factor's.

There needs to be better Environmental Factor's such as Street Light's, better and more Security Cameras and actual Scan Devices put in place at almost every Street Intersection to Serve And Protect; the quantity and quality of these devices to date are so poor one doesn't even need a blind angle to pull off any Crime. If I can just walk in with a gun and a mask then there is no Safety And Security in your Country. A Scan Device and/or Tracking Device can be particularily effective versus Crimes and even Terrorism; now Drones are not failing in War's too. Let them cry Big Brother, Brave New World and New World Order, such already took place with the first Russian Spy Satellite in Orbit. And, anyway, for so-called New World Order here, why is it all still just Not Admissable? According to your paragraphs to paraphrase, '. . . only an Official FBI Agent with a Lawyer after convential means of acquiring witnesses, evidences and proof, and only after the Suspect has reached being an Federal Suspect, can you utilize and/or submit non-convential means, methods, evidences . . . ' And so on . . . Thus, who the Hell are all in violations now and get tossed out of the Court Cases if not all the Court Cases too . . . , 'Oh oh, not another RoboCop Alert!'

Better Home Safety And Security then paranoia which now amount to I'm better off installing my own Internet Security Webcam and uploading to Spy Cloud that way they cannot just rip off the recording too . . .

The original purpose of the Military was to protect the Country from Invader's, not to bomb the shit out of a shit load of sand only. The Military should be seen as Heroes of the Peoples.

Instead of trying to kill 'em all over there we should be strengthening our Defense, Home Defense, our Country Border's, our Internal Affair's and our Homes. By entering a technologically advanced Future starting in the 21st Century we should seriously consider EM Engines, Fission Engines, Laser Pistols, Laser Turrets, EM Shield's, Null Spheres = Null EM Spheres, which is how I already equivalated it, in other words a Laser Military for primarily Defense rather than wasting this much Money, Resources and People per hour in useless Far Out In Left Field Offenses since Nam which to date have gotten us all Squat Squawk Splat! in return, in preparation for the advent of Alien Species in the Future, our Spy Satellites and Space Station's have not a single Laser Weapon of Defense and lame attempts at retorts to EM Pulse Weapon's and ICBM's leads only to Mutual Nuclear Obliteration. Otherwise, goodbye Humanity! Also, if we had any one of the above installations then 911 would have never happened.

Defens ad Absurdum, Poetry ad Infinitum and Poetry ad Infinitum, Defens ad Absurdum.

The real and virtual difference between Private and Public needs to be clarified, strengthened and tighter. It is not legal to spy on your neighbour, especially not on your competition, yet it is a daily, if not hourly, per minute and/or continual, occurence. Due to lack of witnesses, evidence and proof, due to primarily Not Admissable Syndromes, primarily via Internet and unencrypted wireless mediums through the Ether, this is a very serious Modern Symptom which needs to be solved by Specialty Expert's in their Sector's, 'If that's how you want to Spy Hald, be more of a Robin Hood and not a Rebel Teenager, then why don't you zoom in on their keyboard and their password and cause another National Security Issue . . . '

Information Technology in the development of Society's on a Global Scale towards a potential Utopia, the need and want for FREE Information Services is absolutely essential, for without such the Peoples are NOT Informed And Provided for.

Instead of rip-off Software, rip-off Hardware and rip-off Support, Security, Server's, Client's and Services there needs to be a better calculation of the Bell Curve. The Universal Truth of 'If you reduce the cost of your product by $10,00 then you sell 100,000 more copies . . . ' is what I consider great paraphrasing . . . There would be no need, except Industrial Sabotage, for ripping, hacking and cracking to this extent in 2007 alone, now it's 2014 and I look at Security.nl alone and ask myself if everyone needs to buy my own Screw Your Thinking Cap On Device for only 10000000 TE, Augmentation's Cost Extra, or we really have been hit bad by the Enemy and/or Competition . . . If the Peoples' could afford your product then they would buy it . . . Wow, for the first time Pure Logic, though she chooses for only the color, I, myself and me got 100,000+ hits on my websites in 2007 alone, yet not a single donation! It's even worse now in 2014 as they all feed you Pipe Dream's to prevent Mass Panic and our Death Analysis Trend's indicate that it's only going to get worse . . .

Competition is said to stimulate the Economy, but in the cut-throat environment of Free Market's, Free Trades, Free Enterprizes, Free Oppurtunity's and Free Monopoly's it amounts more to non-lethally, and in some cases very lethally, killing your neighbour. If the Country cannot protect its own businesses due to rampant out-of-control Free Competition, Free Sport and Free Noobie Hunting Season then it is no Democracy at all. When International Corporation's can just walk in and wipe everyone else off the map then it is no Democracy at all and most certainly not Unilateral Democracy. When other Country's can do the same then it really is End Of World's . . .

'Space is the final frontier or would it be the brain . . . '

Without Space Exploration then Humanity has no Future whatsoever: Think alone of a whole new World to populate, a Colony Planet, with Infinite Resources, new and advanced Materials, Metal's and Substances. We see now, thanks again to Spy Sat Maya, that there is also not just a couple, with thousands of Galaxy's we got to stop killing each other and Planet Earth or it's going to risk a Mass Extinction Event and cause a Delay Effect for even more than a century . . .

The attraction of encountering new Species and Races, learning from them and trying to categorize and classify them like Darwin and not like your worst version of some horrific extreme neo-commi or neo-nazi, learning we are not the only Living

Being and/or Sentient Species in the Multi-Verses and develop an official system would help eiliminate the stagnant, close-minded, fixed, myopic, Tunnel Vision and Superstition which Humanity suffered from up to and including the 21st Century though Internet is reducing this Blind Domination Effect each and every day as People can inform themselves; with improvements in rates across performance this should accelerate.

'An Ideaology is not a mere Belief . . .'

'Human stood up and bonked his head on a tree branch and the apple fell . . .'

'I think therefore I am Good . . .'

Human called out to GOD but there was no answer, only Infinite Silence, and still is . . .'

'I am what I am what I am what I eat what I say what I do what i yammm . . .'

'I am and I am not a thing . . .'

'To be or not to be a bear with a beard who beared his but or to be but bared or not too bared . . .' You can make your own variations on this one which are extensive but DO NOT touch My Unique Combination. In other words, not to poke too much fun or dis Russia which everyone is doing these days, can I also just 'Rip off your whole Social Democracy', I'm propagating this one that's why see the quotes, or does your Country somehow have Ownership and/or Intellectual Properyt and/or Copyright and/or Register and/or Trademark on it . . . , 'I wonder what we'd be suffering from . . .'

These and many other you have heard multiple times. Since the beginning of Evolution, Human has tried to explain these Mystery's of Existence. To be persecuted for such Religion, Philosophy, Ideaology, Belief and/or Spirituality is utter nonsense. To conduct War, Kill, Murder, Rape and/or Pillage in the name of such is a sin upon GOD himself, not to mention a grievous Insult to the Peoples. Due you really think that Allah wants you to all be Blood Cannibal's? Once again, rather than search for the Truth, Human's pathetic inflated Ego Complex takes over instead, just another wannabe Dictator.

Whether or not Entertainment causes Act's of Violence, whether or not your Film Star, Movie Star, Hollywood Star, Celebrity Star, Celebrity, Star, or you just really like YouTube Star's, is really a God or Goddess, Angel or Demon, Hero or Villain, whether or not frogs, fish, chickens, cats and/or dogs actually fell from the sky may or may not be true . . . the trouble is, you Noobie Human have difficulty distinguishing between Fantasy and Reality . . . you are also a very Superstitious Human . . . A Lizard to a primitive is a Dragon . . . Did I really go biking drunk because of a game . . . Are my 10 Point's for hitting an innocent pedestrian real or virtual . . . Don't believe me, figure it out yourself . . . you cannot help by definition to be in your own Subjective Observation And Experience in an Objective One Big Reality.

'Thank GOD for Science . . .'

Since the 1st Renaissance Science has been the driving force of the Knowledge and Power and Energy of Humanity.

'Knowledge = Power and Power corrupts, Absolute Knowledge = Absolute Power and corrupts absolutely.'

Next to instilling Moral's and Ethic's into the entire Science Community we need to limit the influence of such affluent Scientist's, Laboratory's and Corporation's. With the present rate of Consumption And Pollution there won't be a World left to explore. There will only be one desperate little dingy of an Elite Space Ship to get Off-Planet Pronto!

Science is not per se Anti-GOD, the Church like Nobility has only influence with much less Power and Energy, though not lacking Money, now in the 21st Century, for the argument of GOD is irrefutable: GOD = Creator God = One Big Reality. Otherwise, as I stated in my previous chapter here you are in violation of GOD being Infinite.

Thrice is proof . . . Not in Quantum Science! It's more like if 1000 Empirical White Coat's can prove it, only then it is true . . . otherwise, out of nowhere, again, there comes this highly unpredictable factor which wrecks everyone's calculations . . .

'The World will descend in End Day's into Chaos, Disorder, Terror and Destruction!'

The present Form of Democracy in the early 21st Century on Planet Earth with all its loops, holes, flaws and hypocrisies in legal, judicial, political and financial systems allows for the insurgence and upsurging of Chaos, Disorder, Terrorism and Destruction, hence the need and want for Unilateral Democracy!

If the real and virtual Rules, Regulation's and Laws of the Country are not clarified, strengthend and tightend then it is no surprize the Hun's are at the Gates and the Barbarian's are camping inside the walls.

'The great plans and conclusions of mice and human . . . '

In fact, Unilateral Democracy is the wave of the Future allowing for ALL Thing's and Peoples in Global Civilization's as Global Canadian Citizen's to co-exist in Peace with Freedom and Liberalization.

It can only potentially lead to Utopia in the Far Future when we have many Colony Planet's and not just only in our Solar System since at first oppurtunity we'll also cause a Solar System War . . .

This one Essay is, of course, not the whole story but rather my entire Evolutionary Essays are dedicated to this discussion. See also a later chapter here for a comprehensive definition of a theoretical proposal of Free Democracy. Also, the Philosophical, Spiritual, Religious, Scientific, Political and Economical Community's need to Discuss and Debate it thoroughly and at length, not just in their locations, on TV but also openly on Internet. If you don't mind with all of the hacking, if you're bored give yourself an IT Challenge, I'm not even going to comment on Internet Voting System's, except for one thing: The Internet Popularity Vote cannot fail to Influence Votes.

One of the greatest challenges of the Unilateral Democracy would be actually implementing it and transitionalising systems to it for the survival and progress of Humanity for Century's, if not Millenia, to come . . .

By-Laws

1. Cause And Effect

    A. Nothing is exempt from Cause and Effect. Nothing is Static.

    B. Nothing is Indestructible.

    C. Anything which is not Static or Indestructible can be Moved or Changed.

    D. Thereby, each and every thing can be made to do something and/or to become something.

    E. Thus, everything has a Cause and Effect.

2. Meta-Science

    A. Observation and Experience: One can see the interaction of all Element's and their affect on Life in the cycling process of Season's and the day-to-day growth and decay. See Cause and Effect at multiple sources on Internet. For the rest of the essay I will simply state 'See reference' only.

    B. Science: See Physics. See Law of Conservation of Energy. One form hitting another form and all the Energy, Matter 'hitting', is transferred to the other form. Therefore, one form can effect another form. Thus, form is a thing. See Construction and mobilisation of Energy.

    C. Philosophy: See Eastern. See Greek. See Action and Reaction i.e. primarily Newton's Law applied to Eastern and Western Religion's and Philosophy's. We figure we can also call of Einstein's Law's since and Atomic Bomb has a 10X Factor of a range of magnitudinal force in return. See Not Only Equal In Return.

    If you act then something ensues from such, a reaction.

    If you throw a pebble into an ocean then ripples spread out.

    If you tell someone to Kill, Murder and/or Assassinate someone then a Battle and/or War might ensue on any level. If you tell someone to Destroy something then such can also occur.

    D. Literature: See Poetry Lore—1st Edition—Mass Energy, The Death-Life Conspiracy—Published by AuthorHouse UK: Love Energy, 'While you Breathe . . . forgiveness . . . want.'

    Circular Energy, Death, 'And no other way . . . And come again . . . it's done.'

    Circular Energy, Death, 'Dreadful . . . Or would you rather . . . own happiness?'

    Circular Energy, Life, 'Puzzlement . . . Just stop kicking . . . your feet.'

Circular Energy, Death, 'So now that I sit . . . what causes it? Slouched over and down.'

Circular Energy, Death, 'Competence . . . Move by . . . Never to . . . That by one . . . will live.'

Circular Energy, Death, 'And I Live . . . If the Effort . . . Breath of . . . With each . . . motion.'

See Poetry Lore—1st Edition—The Power Of Release—Published by AuthorHouse UK: All Things, 'All Things . . . Forces Enacting . . . This is a . . . people.'

On The Same Street, 'Isolated . . . No . . . Cause . . . Working . . . If . . . Safely.'

Collide-A-Scope-Worlds, 'Come on . . . It is . . . With good . . . Forever.'

You And I, 'This Fucking . . . society.'

See Poetry Lore—1st Edition—Self-Consciousness, Law Of Unification (all bodies)—Published by AuthorHouse UK: Self-Consciousness, 'My guiding . . . What to . . . My duty . . . I will . . . Poet!'

You Are A, 'Hello Non-Sequitor . . . Clearly . . . Repetitious . . . Z.'

Superstition, 'Human, you . . . Superstition . . . So it might . . . Here.'

A Little Poem For You, 'Dear Father . . . You have . . . In fact . . . peacefully.'

Cause and Effect, 'Life is . . . Death is . . . This . . . Millions . . . Is . . . Light.'

Heart, 'I have . . . Letting . . . I say . . . Then what . . . Beyond . . . Thank You!'

Evolution To Poetry Lore—1st Edition—Self-Consciousness, Law Of Unification (all bodies)—Published by AuthorHouse UK, 'Oh . . . Amoeba . . . A Random . . . Human.'

You Have To Fight For Freedom, 'I am . . . To Survive . . . All . . . Universe.'

See The Black Dungeon Doorway—3rd Edition—Published by AuthorHouse UK: 'When you said black . . . Orthe: There are countless . . . I . . . '

'Cool. Did we or . . ."Wow. I didn't . . ."Action and . . . wanted to.'

'It's time to . . ."Everything . . ."Water is . . . See yah.'

See Evolutionary Essays (this official formal and informal thesis extending from York University, Toronto, Canada. See the latest version of my curriculum vitae at one of my two websites: Silverlingo.com and/or Planesofexistence.eu):

Change, 'This . . . Change . . . To Respond . . . If . . . Population.'

The Conundrum Solution, 'The Trunk . . . will Cease.'

Change Is The Only Constant, 'We . . . reality.'

Cause And Effect, 'No Wonder Now . . . Grain of Sand.'

Each reference sheds light on the nature of Cause and Effect. Though the references are found within other frameworks i.e. Superstition, they provide the key phrases and/or lines where the exact nature of Cause and Effect are expanded upon. Here, I am providing argumentation, examples, references and evidence which will hopefully lead to proving the Truth of the existence of these By-Laws. Though I certainly do not claim absolutism over the nature of one thing or another, such as even Cause and Effect, since Quantum Science is showing that it's not as simple as Newton and even Einstein thought through non-local causality and confusions in Chaos Theory of so-called Random Event's which are still inexplicable and the still highly debatable Karma Theory's, I would not want my previous works to go unread and/or not cross-referenced which I have now finally been able to develop up to a certain tier of existence and I consider them to be a vital and recommended reading, especially if you are going to continue reading this one. If you already have, then hey, read 'em again! You might not have the latest versions which are only at my two websites; try not get them via via because of a not-so-new IT Phenomena called I Cannot Give Them My IP Address since your little brother or sister will do some Editting On The Fly. Once again, a single 'not' in that sentence with all their Not Not World's these days and you will cause another National Insult. Thank You (heh heh heh) . . .

Also, with these proffered references and cross-references, light is also shed in return to the works themself.

3. Do What You Want As Long As You Harm No One

This statement originates from Celticism/Paganism/Wicca.

Being the direct offshoot of Cause and Effect it is the primary By-Law one Human must govern their causes by.

Nature has everything you want. All of our needs, our will to survivie, all of our cravings, our deep innermost desires, are found within Nature or One Big Reality.

Nature = One Big Reality. This is why I still think the term 'man-made' is an oxymoron since Human is A = A First Instance created by Nature.

Here though we, of course, have the GOD Principle which is still a Moot Point, however I still have no problem even making this equation: Nature = One Big Reality = GOD. This is the millennia long confusion in terminology and cross-language barriers, everyone intuitively associates the word Nature with only our Environment on Planet Earth due to overly excessive Green Liberal Issues and their Opposite Enemy Industry Conservative. Well, I also have no problem here adding to the equation: Universe Environment = Nature = One Big Reality = GOD. Once again, if GOD is not also in and/or one with such then each of you are actually in violation of an Infinite GOD. See previous essay here.

True Love is based on the premise of Nature.

By working on this one By-Law, The One By-Law, we can become whole and free. We will be able to become completely self-sustaining, self-sufficient, self-satisfying and also still interact with others without being a Life Killer and/or Blood Cannibal. And most importantly, we will be capable of doing anything, of fulfilling every desire. Effectively this cannot fail to also lead to The Holy Grail and True Immortatility which is most certainly not your blow Hollywood Media Puppet with Rich And Famous Hit's only . . .

Your are allowed by Nature do Do What You Want As Long As You Harm No One.

This may seem by now a little tedious but consider this: It is The One By-Law. And, consider this: This One By-Law encompasses All Thing's, Group, You, Me and Nature.

Why would Nature invent any more? Nature is the Big One. Relative to our systems this One is vastly more congruent, simple, and real. Indeed, this By-Law is the distillation of a vast quantity and quality of Object's, Subject's, Values and Variables.

Here is another perfect example of confusion in terminologies: Do you mean ONE, One, Only One, One Only, ONLY One, One ONLY, ONLY ONE, ONE ONLY, Big One, One, The One or any other attempt to figure out GOD . . . Where it is placed in object context oriented languages from English to PHP is also very important for the actual meaning over the millennia could have been easily lost from A $\rightarrow$ Z and is now unfortunately causing severe conflicts leading to violent and bloody Debates, Battles and War's on Planet Earth in beginning of 21st Century, and even to date for millennia long as being *the* One Cause Of Conflict, not just Faction Fraction Fractal Fucture Display.

And, of course, this is not Black And White Complex. A Tree can and should be considered a Living Being, an One, an Individual.

There are very easy ways to prevent the razing and slaughter of such Trees, Plant's and Animal's in such Jungles and Forest's. See previous essay here.

Life is a process of development. What counts and what actually allows a One to achieve what they want is the continual work towards this pure and absolute One Law.

You do the best you can. K.I.S.S. Uncategorically, I am certainly by no means, whatsoever, in any way, all ways, exempt from this! (Whew.) I am Proud (foot) to say, "I am Human!", too. (Whew again, I can almost feel the Hell Fire a coming again . . .)

Due Note: 'If you ever doubt the Path of Truth, Just look at the rewards . . . '

Nicht Ontheil Veritas!

4. Meta-Science

A. Observation and Experience: Each and every moment in the momentum of our lives in the Present we can see the obvious result of things done in the Past, therefore the Future also benefits or suffers from such. The acts we make as an Artist in the Art in the present momentum decide and determine each of our Futures for at some Intersection Point in our Timeline we may or may not Interact with each other. We gain friends if we are friendly to others. We each have Degrees Of Good, Neutral and/or Evil, and also Degrees Of Ally, Neutrality and/or Enemy. This why I keep saying we have a Near-Impossible Paradox in Karma Theory for the only way not to cause more Karma is to Not Interact therefore is to Not Exist. This is like saying Nirvana = Infinity = 0 which is erroneous for in Mathematic's we have Infinity = ∞. This also looks like the same confusion with Nothingness, Nothing and Ether since now thanks to Maya Spy Sat again what is in between each Galaxy? Yes, Ether. Yet, to them, never forget each of your Mystic's in your Religion next to having an Oath Of Secrecy had to describe everything in Symbological Terminology; I'm not saying you're per se incorrect but in that millennia and century and year of 550 BCE without any Modern Science And Technology with the necessary Modern Devices it could very easily be misinterpreted.

This is the greatest error and flaw to date by Human's looking only and continuously through their Subject Reality Glasses.

'Also do not presume to know the Will of GOD!'

'DO NOT presume to know the Will of my God and/or Goddess!'

'Do Not Interact with me unless you want to suffer the Hell Fire of my God Of Hell Fire you fucking goddamned Twat Nüber Kanker Apschen Mensch!'

Like I said in previous essays here, don't mind my B—Stupid Violent Black Humor to kill the long dry monotony of most philosophical essays to date written by highly over-complicated anal intellectuals who don't even dare say 'shit' word. Honestly, I can't help cracking up laughing, now and zen, it's like they have No Sense Of Humor and in any case here's My Scottish Up Yours . . .

B. Science: Physics. There is an equal and opposite directional reaction to every force applied to paraphrase Newton's Law, 'Why you hit a wall, the wall, does punch back, equally . . . ' says your average Aries who seems to go on for even years and years out of some Stupid Sense Of Stubborness and then asks him and/or herself why he/she now has a Big Bruise On Bonker and the Scottish Wall Is Still Silent And Standing. Equal In Return. There is a 10X Factor ranged magnitudinal force in return to paraphrase Einstein's Law's, 'Why when I kick a dog does a horse show up and kick me in the head 360 km/h into a tree . . . ' Not Equal In Return.

C. Philosophy: Celticism. Paganism. Originated from the Pagan Tradition of the European Continent and British Isles with Druid's and Priestesses very wrongfully branded by Roman's with the labels and images of Heathen's and Witches dating back to the Celtic Era from Proto-Celtic (c. 1200 BCE) to Hallstatt Culture (c. 800-500 BCE) to La Tène (c. 450 BCE up to the Roman conquest) to Gallic Invasion of Balkan's (c. 279 BCE) to Central Anatolia ruled by Galatians (c. 3rd Century BCE) to Celtic Culture and Insular Celtic Culture (c. mid 1st Century CE) which had become restricted to Ireland to Wales, Scotland, Cornwall, the Isle of Man and Brittany enjoying a cohesive cultural entity (5th to 8th Century's CE) to Continental Celtic Languages no longer in wide use (6th Century) to Celtic Christianity or Insular Christianity in Early Middle Ages (5th to 10th Century) to a uniquely Irish penitential system adopted as Universal Suffrage of the Church by the Fourth Lateran Council of 1215 to almost a whole millennia of suffering and punishment through suppression and repression in Middle Ages (c. 5th to 15th

Century CE) and Feudal Ages (c. 5th to 15th Century CE) by Inquisitor's and Tax Collector's to the Reformation and the Renaissance after the outstanding period of Dafydd ap Gwilym and his followers in the cywydd where arose for a short time a school of literary formalists, the chief of these was Dafydd ab Edmwnd whose poetic heirs were Tudur Aled (c. died 1526 CE) and Gutun Owain (c. 1460-1500 CE) coinciding with European Renaissance (14th to 17th Century's) to a Modern Celtic autonomous identity in Western Europe following the identification of the native peoples of the Atlantic fringe as Celt's by Edward Lhuyd in the 18th Century to Ethnic Nationalism particularily within the United Kingdom of Great Britain and Ireland where the Irish Home Rule Movement resulted in the secession of the Irish Free State in 1922 (late 19th Century) to the last 200 years where the Celtic Language has diverged and become three Languages: Irish Gaelic, Scots Gaelic and Manx. As a Culture Group we encourage you to use the present term and classification: Gaelic. See Wikipedia.com in English for all of these, except the one for Reformation and Renaissance see Encyclopedia Britannica.com .

Paganism is a Nature Religion of Duality believing in the manifestation of the Man and Woman Power and Energy on all levels, the ultimate Woman, being a Goddess, the ultimate Man being a God.

See also Eastern Religion's and Philosophy's. In case it was unclear so far I include Mysticism within these two utilizing The Land Of Andor principle. See previous chapter here.

If you act like an Artist in the Art, Karmic Particles attach themselves to you, thereby giving you Soul, Spirit, Mind and Body bondage with Degrees Of Power and Energy. If you make a Not Violent Act then you gain positive Karma. If you make a Violent Act then you gain negative Karma. If you make a Neutral Act then you gain neutral Karma. The present confusion in the Topic, like my expression in a previous paragraph here, is how do we really unbind or dissolve various Types Of Karma and why does the form of the manifestation of Karma over Timelines not always be the same. In other words, why does the Cause not line up with the Effect in both quantity and/or quality. This is one of the mysteries of Buddhism, Zen Buddhism, Confucianism and others which as a Buddhist once said to me, 'The only way you can know or be convinced of something is to go and find out yourself.' I also like 'The only way is your own will' and 'If there is no will then there is no way'. See Karma and Reincarnation in many Religion's and Philosophy's, and as far as I am concerned Science backs it up to with Newton's Law, Einstein's Law's and Law Of Conservation Of Energy.

D. Literature: See Poetry Lore—1st Edition—Mass Energy, The Death-Life Conspiracy—Published by AuthorHouse UK: Prologue, 'Do What You Want As Long As You Harm No One'

WTLOLAGAFE, 'That day the blood . . . no reprieve.'

Love Energy, 'Romance is a . . . fleeing . . . Take heed . . . beasts.'

Circular Energy, Life, 'Oh I . . . Busses . . . In the turning . . . Who cares?'

Circular Energy, Death, 'What a waste! . . . I'll take . . . BLAM!! . . . See we . . . again . . .'

Circular Energy, Death, 'How Dare . . . yes! It is . . . Creatures! . . . Brother . . . Hide.'

Circular Energy, Life, 'Dark enshrouded . . . Sin to . . . These are . . . know it.'

Circular Energy, Life, 'Live of Life . . . Do not . . . Would you . . . It is rotting . . . die.'

Circular Energy, Life, 'This world . . . So I kill . . . You can't . . . It's only . . . here.'

See Poetry Lore—1st Edition—The Power Of Release—Published by AuthorHouse UK: Power Trip, 'Now it's Time . . . Release . . . We will . . . want.'

Final Tribute . . . 'The unconscious . . . Oh why . . . So let's be . . . breathe.'

Collide-A-Scope-Worlds, 'Come . . . It is so . . . You have to . . . It is fun . . .'

You Will, 'The life . . . You took . . . You believe . . . If you ever . . . rewards.'

See Poetry Lore—1st Edition—Self-Consciousness, Law Of Unification (all bodies)—Published by AuthorHouse UK: Anger, 'To be consumed . . . My opponent . . . The, to . . . Realizing . . . end.'

Self-Efficiency, 'Yea must . . . Don't . . . Then you. You will be . . . you.'

HITDOYF1, 'To die, to die . . . I would . . . How can you . . . now.'

Quote: Albert Einstein, 'Being Vegetarian . . . survival on Earth.'

Quote: Golden Verses, Pythagoreans, 'Pay honour . . . a mortal man no more.'

Two Scientist's Outside . . . , 'Oh Despair . . . Ahhh . . . Yes . . . Well . . . Yes.'

See The Black Dungeon Doorway—3rd Edition—Published by AuthorHouse UK: 'Man, we are violent . . . Aera continues . . . underlying law.'

'What is Life? . . . What is wrong . . . He ignores . . . Orthe . . . his feet.'

'You Can Do . . . The Do What You . . . to yourselves.'

Evolutionary Essays (this formal and informal thesis): 'Aware . . . Go . . . Do not harm . . . Since . . . no longer Violence.'

'Imagine . . . So do you . . . Am I supposed . . . If you are . . . have.'

Your Guilty Conscience: 'Onto . . . Stop . . . Get paid . . . satisfied.'

Polaris-Politic's II: 'Our present . . . We must realize . . .'

Human Sanity: 'We . . . To go . . . Thus far Humankind . . . Now . . . Society.'

Health: 'Let . . . When . . . They . . . If . . . Now . . . We . . . Diet.'

Cause and Effect: 'No wonder . . . Karma . . . All . . . Love.'

Karma and Reincarnation: 'Energy Matrix . . . is maintained . . . look-alikes . . .'

After Life Party: 'I don't think . . . I want to be . . . stuck in that groove . . . dance mode . . . techno mode . . . metal mode . . . Hell Mode . . . for that long . . .'

5. All Things To Balance

A. Every little bit of Power and Energy works so as not to work, to reach a position of rest. A position of rest is one of Balance where the least quantity and quality of forces must be exerted so as to maintain its state, structure and/or position. Thus, Symmetry, Balance just like a LASER.

In fact, perfect Balance, perfect Symmetry, a Perfect Sphere results in Zero expenditure of Power and Energy. Like I said in a previous chapter always leave something to the Mathematic's and Physic's Expert's, thus good luck with this one. Technically speaking it could also be Infinite Power And Energy through A. Expansion Theory and B. Equal Input And Output and C. One Big Reality = Infinite GOD. I'm starting to think and lean though towards even No Expansion Theory due to the problem it is causing with Finite, Near-Infinite and Infinite Issues. Also why is Andromeda and Milky Way Galaxy actually moving towards each other on a collision course as two of the biggest Cannibal Galaxy's, I guess the microcosm really does reflect macrocosm and not per se vice versa, and regardless if it takes zillions of millennia what kind of Mad GOD would destroy everything through Black Hole Theory and Big Bang Theory devouring it all to One Point Of Near-Infinite Singularity and then exploding again . . . See the problem here, would that not have to be One Point Of Infinite Singularity?

6. Meta-Science

A. Observation and Experience: It is obvious to see our Culture possessed with the pursuit of Comfort And Luxury. Luxury is primarily associated with rest, yet we do so much in American Dream Complex which practically annihilates the desired effects that I'm wondering if everyone has lost their Cookie Banger, still though it does not fail in the Work Rest Play Theory. People move continuously about to maintain their level of comfort just like we War so much for our Peace. And if one more says Nature Of The Beast then I am going to repeat again it's more like Nature Of Lucifer and/or Nature Of Satan which everyone keeps forgetting asking themselve how cruel can GOD be . . .

B. Science: Physics: The Law Of Conservation Of Energy states the quantity and quality of Power and Energy is constant. A form has a fluctuating quantity and quality of Matter, Energy and/or Power which when the Particles, Lines Of Energy, Lines Of Power are disrupted and/or broken is transferred and/or transformed into another Form, usually the one in closest proximity. It will move to Conserve its Energy and to keep its Power and Energy Constant. See also Table Of Element's in Chemistry and Biology.

C. Philosophy: Eastern: I Ching. All Power's and Energy's are composed of Yin/Yang. The Woman balances out the Man. The opposite to Pain is Pleasure. The solid line is Yang and the not solid line is Yin.

D. Literature: See Poetry Lore—1st Edition—Mass Energy, The Death-Life Conspiracy—Published by AuthorHouse UK: Spiracle Energy, Spiral Motion, 'Up and Down . . . Round and . . . again.'

Spiracle Energy, Body Spiral, 'On white plane . . . Welcome to . . . Entropy . . . . same.'

Spiracle Energy, Love Spiral, 'What brings . . . My quelling . . . Now I call . . . LOVE?'

Spiracle Energy, Mind Spiral, 'My body . . . No escape . . . Contradiction . . . On.'

Circular Energy, Life, 'This is . . . The dying . . . To birth . . . beginning.'

Circular Energy, Life, 'Oh I wonder . . . All I want . . . I get off . . . Why . . . cares?'

Circular Energy, Death, 'I do not . . . And that is why . . . No more . . . again.'

Circular Energy, Death, 'Competence . . . Move by . . . Never . . . All . . . You are . . . live.'

See Poetry Lore—1st Edition—The Power Of Release—Published by AuthorHouse UK: Power Trip, 'I trust . . . Now it's . . . Release . . . We will now . . . want.'

All Things, 'All things . . . Balance points . . . This is a . . . Attraction.'

Progression, 'Each Plane . . . The way . . . From one Cycle . . . By . . . Plane.'

Collide-A-Scope-Worlds, 'Come . . . Worshipping . . . There is . . . Forever.'

Your Land, 'You go to . . . Come to . . . Hit the . . . Step on . . . Now . . . sky.'

'Diamond . . . With our . . . So let's . . . body.'

Coming To . . . 'Beautiful . . . Finely . . . Balance . . . You . . . In the fusion . . . '

Gates Of Dawn, 'Here we go . . . We're . . . Back . . . In time . . . Dawn.'

See Poetry Lore—1st Edition—Self-Consciousness, Law Of Unification (all bodies)—Published by AuthorHouse UK: Self-consciousness, 'My . . . And all I . . . Allow . . . I will . . . Truth . . . Poet!'

Friendship, 'Journeying, as one . . . To hold . . . Having . . . Abound . . . Love.'

A Little Poem For You, 'Dear Father . . . Give . . . The Plateau . . . Go . . . '

Cause And Effect, 'Life . . . Death . . . Do not . . . The trick . . . Light.'

Self-Efficiency, 'Yea . . . Don't . . . Then . . . You will . . . you.'

Hell Is . . . , 'And reach . . . I would . . . We have passed . . . Energy . . . now.'

The Secret . . . , 'Consider it as a whole . . . Now Here.'

Truth, 'Yes . . . There is . . . Average . . . True . . . You are well come.'

Truth Is Commonality, 'All things . . . Earth . . . Holding . . . In . . . Eyes.'

Two Scientists, 'Oh Despair . . . Yes. Restrain . . . Yes.'

You Have To . . . , 'I am Back! . . . Strive Up! . . . Malnutrition . . . Universe.'

Freedom, 'The rocks . . . around spread . . . their very edge . . . beyond . . . wings . . .'

See The Black Dungeon Doorway—3rd Edition—Published by AuthorHouse UK: 'So I'm . . . He stands . . . I will . . . I can hear . . . consciousness.'

'Wow . . . Action and . . . Orthe about . . . surface.'

'It's time to . . . See yah.'

See Apotheum Colluseum—The Ultimate InterActive™ Game—2nd Edition: Elemental Forces, 'The Elements . . . They are your . . . Fate.'

KP Balance, 'This is the Balance . . . You Choose . . . goes.'

Action Form's, 'Stability Action force, Breaking Action Force'

POP = Known—Unknown, 'As the Unknown . . . POP Chart . . . Slave . . . Master.'
Hold It Together, 'By Human Plane . . . 7 Power . . . It is . . . that.'
Modifier's, 'EF's . . . To understand . . . As Result . . . Polarity.'
PAHHF Roll, 'You will . . . If in cases . . . This is allowed . . . If you Succeed . . . '

Place, 'Each Place . . . REP . . . Each Place . . . Energy . . . degree.'
Evolutionary Essays (this informal and formal thesis): The whole thing.

7. Repulsion and Attraction

A. Every Object and Subject in Nature is in connection to every other Object and Subject, whether directly or indirectly, by Range. These atoms are pulled towards each other or pushed away, resulting in various larger sized Object's and Subject's. Each atom is almost indestructible so this works.

Also ensuing, are further motions, notions, oceans, and with furthering of these, more fusions.

Overall is perceived an emulsion of Collide-A-Scope-Worlds. This is more evident in Living Beings, One's, Object's and Subject's with Electricity running through them than Non-Living Beings, thus for some more B—Stupid Violent Black Humor obviously not only someone being electrocuted; it's still remarkable the statistics on how many people have survived being hit by lightning and other Kitchen Act's.

After all, Living Beings move around more . . . are more active and posses more Power and Energy, and of course not only The Pope.

8. Meta-Science

A. Observation and Experience

We fall to Mother Earth. The Earth resists our intrusion. People of similar orientation find they spend time with each other. Likes, associates attract, not only opposites, Equal In Return also has the opposite Not Equal In Return, which is a result of the defining of two poles in Science, but quite often likeness. We each can be placed under Like Group AND The Fact You Lack It You Are Attracted. So, the attraction of opposites results in an exploration of the BAF Unknown AND the SAF Known. See Apotheum Colluseum—The Ultimate InterActive™ Game—2nd Edition for these acronyms.

B. Science: Physic's: Electro-Magnetism.

All Body's have an Electro-Magnetic Field (EM Field or EMF) which attracts Particles, Lines Of Energy and Lines Of Power to them. Living Being's, Object's and Subject's with Electricity pouring through have a stronger EM Field based on the quantity and quality of juice moving through them. I prefer Beer, Wine and/or Alchohol and don't need to hear one more Immi Noob objecting to our own Ancient Habit's And Tradition in our European Cultures and Colony's. Object's and Subject's with an aligned EM Field flow in the same direction and move towards each other when parallel, like a Magnet with iron filings. This is symbolically defined in Mathematic's and Physic's with + and—signs for the Magnetic Poles, like Planet Earth and every other Celestial Body also has. What it actually means is, like the Black & White Pen Art in Poetry Lore—1st Edition—Published by AuthorHouse UK) is the direction of the Lines Of Energy and/or Lines Of Power which are coursing through the Timelines of the EM Field's. What I recently read in a Scientific American (Aug 2013, Meinard Kuhlmann) which is related to this Topic, and for the first time yesterday (09-03-2014) I comprehend Vector Graphic's (as in i.e. Adobe Illustrator and a number of 3D Games), the EM Field has to have a rate of state with a path, like a hyperbole and trajectories, and requires structure, like WWW or the brain's neural network, thus not denying dimensions and planes, like low polygon or high polygon 3D Graphic's, thus as he states 'The need to apply the quantum field to the state vector . . . ' What is funny and coincidental is that just before I read this article in the park I was busy with a difficult Burfday Challenge to Optimalize Win7 and was cracking jokes in our 3D Chat Environment about State Of System (SOS) where you have to make a valid or good Restore Point.

C. Philosophy: Eastern: Taoism: The 100th Monkey Effect or 101th Monkey Effect or 13th Monkey Effect or 12th Monkey Effect. If you cause motion here it can manifest and multiply elsewhere. Do not do it with me though, it's called blatantly ripping someone off through Lie Thru Your Typhus and acting like it's AI and Lies And Excuses and Lie Cheat Steal Hack and plagiarism, copyright, Intellectual Property and Ownership Violation's, I type, scan and upload everything I make soon enough to Family, Friend's, my two websites on Internet and/or my publisher, I also surf enough and will find out and sue you for at a min of $2,000,000 netto, some say my own Personal Net Worth is that amount and my Total Net Worth, not including Prototypes to come, is $200,000,000, it's also not recommended to do with anyone else either, if your very own find out then you'll have the Worst Type Of Pie On Your Face, and when your own leave you then that's the worst part . . .

Western: Paganism: If you energize your gonads you can create Great Magic, which is Quantum Energy. Sex is a perfect example of repulsion and attraction, the Magic Being, the Moon Child, however just to kill the stereotype again, there is obviously not just the Full Moon for us but the Sun, the Sky's and the Star's.

D. Literature. See Poetry Lore—1st Edition—Mass Energy, The Death-Life Conspiracy—Published by AuthorHouse UK: WTLOLAGAFE, 'So is . . . Who's to keep . . . On the . . . But an . . . '

Through Time, 'Look through . . . In the region . . . You have . . . As Gravity . . .
'

Spiracle Energy, multiple poems.
Love Energy, 'While your Breathe . . . Embrace . . . Flowing . . . beasts.'
Circular Energy, Life, 'And I Live . . . If the effort . . . Breath of . . . motion.'
The Things I Like, 'Terrific . . . The happy . . . Having a . . . Playing . . . goal.'
See Poetry Lore—1st Edition—The Power Of Release—Published by AuthorHouse UK:
Thank You Mom, 'This is . . . to.'
All the rest.
See Poetry Lore—1st Edition—Self-Consciousness, Law Of Unification (all bodies)—Published by AuthorHouse UK: All poems.
See The Black Dungeon Doorway—3rd Edition—Published by AuthorHouse UK: Each page.
See Apotheum Colluseum—The Ultimate InterActive™ Game—2nd Edition: In its entirety.
See Evolutionary Essays (this formal and/or informal thesis): Each chapter.

9. All Thing's Are Physical

A. All Thing's Are Real.
All Truth can be understood through the Physical, the real. There are many levels of physical substance, Matter, which as particles move through each other. Things are in Every thing, Everything is Nature. Nature = GOD.
Only No Thing is external to Every Thing.
Nothing ≠ Everything.
Object's and Subject's are Real and/or Virtual. This is a necessary lenience in definitions which, once again, due to terminologies and cross-language barriers has also caused millennia long confusion about Existence; eachObject and Subject does exist. However, they exist in Dimension's in Planes Of Existence which are continuously fluctuating between Real and/or Virtual State Vector's.

10. Meta-Science

A. Observation and Experience
You can touch everything in Nature. A field of substance is the air. You even breath them.
B. Science: Physics: Einstein: $E = MC^2$. Einstein's equation shows Matter is not fixed as Newton thought, it shows the underlying motion of Matter, Energy, is in fact more than Matter, simply smaller and faster. To quote: 'We may therefore regard matter as being constituted by the regions of space in which the field is

extremely intense. There is no place in this new kind of physics both for the field and matter, for the field is the only reality.'

So, Reality is a fluctuating EM Field of various densities.

This can be seen with Earth, Water, Air, Fire and Ether the 5 States of Matter. Ether is hardly associated with Matter and erroneously so as I described in a previous paragraph here.

C. Philosophy: Western: Meta-Science. With Truth as its guiding 2nd By-Law for the purpose of becoming One with Nature for we can only define things by their opposites and Truth Is Commonality. See previous chapter here.

There used to be only Science which argued all things are real, one of Science's highest virtues. Now Here there is Meta-Science. Here Now, "Here Now, Here Now, a new Voice has hit the land!" And not only her lamb . . .

This has been the saddest things of all Philosophy's: Their denial of the Real, Solid, Physical Existence of Nature. To even argue Perception is Illusion is misleading; there is simply more to it which you can't sense, yet . . . you CAN ONLY look through your own Subjective Reality Sunglasses. This is a Truth, if not just a fact, those who dare try and contradict it are the worst Nietzchean types. Another Point Of Debate I want to bring up here which I have heard, read and encountered multiple times is in our Quest for The Holy Grail leading to mythical Immortality which only at a min of Demi-God Heroes possess, like Hercules, I am of the opinion, and also Theosophy backs this one up, that you cannot become GOD, only at best One With The Universe, otherwise you are in violation. See Greek Philosophy, Theosophy and many other Religion's and Philosophy's.

D. Literature. See Poetry Lore—1st Edition—Mass Energy, The Death-Life Conspiracy—Published by AuthorHouse UK: Time Energy, The Clock, 'Time only . . . Time does not . . . It gives . . . it.'

Love Energy, 'While you Breathe . . . Embrace the . . . Flowing in . . . beasts.'

Circular Energy, Death, 'They . . . Swim in . . . While . . . In the sea . . . Find your Body.'

Circular Energy, Death, 'I do not Strive . . . I fell frail . . . This is why . . . again . . .'

Circular Energy, Life, 'Of I . . . Round . . . Arrggh . . . Blllurp! . . . Oh I . . . HAH HAH!'

Circular Energy, Death, 'So now . . . States . . . What makes . . . What is . . . What are . . . Is it the . . .'

Circular Energy, Life, 'Love of Life . . . The seed . . . Do not . . . The Flesh . . . Bacteria . . . die.'

Circular Energy, Death, 'Competence . . . Rice the . . . Break new . . . New tiers . . . That by . . . live.'

Circular Energy, Life, 'Terrific . . . Drinking . . . Wonderful . . . Squirrels . . . Having . . . Playing.'

See Poetry Lore—1st Edition—The Power Of Release—Published by AuthorHouse UK

In Return . . . , 'Something nice . . . Whistle . . . creations.'

Your Life Fire, 'Fight for Life . . . Fight for air . . . Fight for high . . . can.'

All Things, 'All things . . . Are Physical . . . There are No . . . This is a . . . people.'

Travelling Song, 'Dream . . . Release those . . . Enter . . . Face . . . Let your . . . it.'

Collide-A-Scope-Worlds, 'Come on . . . Forever.'

Your Land, 'You got to . . . Come . . . Let Gravity . . . Hit the land . . . Step . . . Now . . . sky.'

Knight In Shining Armour, 'The Knight . . . Whether . . . They . . . Shed . . . Avalon.'

'And if you . . . On this . . . With our . . . So let's . . . You don't . . . body.'

Coming to . . . , 'Beautiful . . . Finely . . . Pattern . . . You couldn't . . . In this . . . '

Gates Of Dawn, 'Here we go . . . I like this . . . We're going . . . And now . . . Dawn.'

See Poetry Lore—1st Edition—Self-Consciousness, Law Of Unification (all bodies)—Published by AuthorHouse UK: Superstition, 'Human, you . . . Like . . . One must . . . Based on . . . There is . . . So it . . . Now Here.'

A Little Poem For You, 'Dear . . . With . . . The Plateau . . . Not to . . . The body is . . . '

The Secret To . . . , 'Consider . . . Now Here.'

Truth, 'Yes, Truth . . . Are you denying . . . What is . . . Truth . . . come.'

Embetterment, 'In . . . Breathing . . . No matter . . . And grasp . . . No matter . . . Back.'

Truth Is Commonality, 'All things discovered . . . Are Elementary . . . Eyes . . . '

Quote: Albert Einstein, 'Being Vegetarian . . . survival on Earth.'

Evolution To Self-Consciousness, 'Oh . . . And, they . . . From . . . A Random . . . Human.'

Quote: Golden Verses Pythagoreans, 'Thy belly . . . But give it . . . Eat not . . . more.'

You Have To Fight For Freedom, 'I am Back! . . . Push up . . . Energy . . . Malnutrition . . . Universe.'

Freedom, 'The rocks grow easier . . . with perfect . . . their very edge . . . wings . . . '

See The Black Dungeon Doorway—3rd Edition—Published by AuthorHouse UK: 'In that in between . . . predictable.'

'Losing everything . . . sensations? . . . Boo hoo . . . After all . . . hallway.'

'Ahhhhh . . . Ahhhhh . . . Ahhhhh . . . Ahhhh . . . Cut Back . . . No . . . After . . . damned.'

'It's time to . . . Everything is a thing . . . Those are your . . . yah.'

'Pyre, Aera . . . This is Paradise . . . forever.'

See Apotheum Colluseum—The Ultimate InterActive™ Game—2nd Edition: Its Entirety.

See Evolutionary Essays (this informal and/or formal thesis): Its Entirety.

See Multi-Verses, Universes and all Galaxy's.

See its Entirety, One Big Reality.

Form's Of Motion

There are 3 primary Form's of Motion.

1. The Spiral

    A. The Spiral is the 1st Form of Motion.
This is the Great Form, Unconsciousness, Death. If you notice, it has no beginning or end. The Spiral pulls one point to another point, both points indefinite, thus the Form of Death. It achieves nothing, fast. See Spiral Galaxy's.

    B. Meta-Science: Observation and Experience: Anything falling will spiral to the ground i.e. a leaf falling to the ground.

    C. Science: Physics: The Right Hand Rule.
The motion of particles around a Line Of Force. You put your right hand into the Thumbs Up' position. Your thumb shows the direction of flow of Power and Energy. Your fingers show the clockwise spiral around your thumb; the opposite direction is counter-clockwise.
Examples of these are electrical wires and the most important: Magnet's.
Astronomy: A Galaxy is composed of at a min of two opposite spirals: 'Spiral galaxies come in a wide variety of shapes. Roughly 60 percent of spiral galaxies contain multiple arms, while another 10 percent have only two. Approximately 30 percent of spiral galaxies lack well-defined arms, as their features have faded over time.' Source: Space.com/22382-spiral-galaxy .
Medical Science: DNA: The double helix of the chromosones also follows this.

    D. Philosophy: Paganism: The two identical, however opposite in direction, primal forces of the Man and Woman are represented by two opposing spirals. These two compose the Universe. This is represented and portrayed in pretty much all Celtic Art and Artifact's. See Wikipedia.com in English.
Theosophy, Greek and Eastern Indian Religion's and Philosophy's: Hermetic Snakes. The two intertwined sticks around Hermes staff, the hippocratic symbol of Health, are in the double spiral form. This represents also Pranic Energy with Chakra Point's amongst other things.
Nicht Ontheil Hermes! Nicht Ontheil Hippocrates! Nicht Ontheil Hygiene! Nicht Ontheil Tirthankaras!

    E. Literature: See Poetry Lore—1st Edition—Mass Energy, The Death-Life Conspiracy—Published by AuthorHouse UK: Spiracle Energy, each poem.

See Poetry Lore—1st Edition—The Power Of Release—Published by AuthorHouse UK: Power Trip, 'I trust my flow . . . And death . . . occur.'

Progression, 'Spiralling . . . To Connect . . . The way . . . To Travel . . . Plane.'

Travelling Song, 'Dream . . . Enter . . . Face . . . What are . . . Let your . . . it.'

Final Tribute . . . , 'The unconscious . . . I breathe.'

Collide-A-Scope-Worlds, 'Come on . . . In their . . . When things . . . Forever.'

'And if you . . . With our . . . So let's . . . Spiralling . . . You . . . body.'

Gates Of Dawn, 'Here we go . . . I like . . . We're . . . In time . . . And now . . . Dawn.'

See Poetry Lore—1st Edition—Self-Consciousness, Law Of Unification (all bodies)—Published by AuthorHouse UK: Anger, 'To be . . . My opponent . . . Then . . . Realizing . . . And those . . . end.'

You Are A, 'Hello Non-Sequitor . . . Playing with the spiral . . . Z.'

A Little Poem For You, 'Dear . . . The Plateau . . . You cannot . . . The body . . . '

Cause and Effect, 'Life is . . . Death is . . . As you . . . Do not . . . So . . . Light.'

The Pathway To All Realities, 'My motivation . . . To All Realities.'

Self-Efficiency, 'Yea must . . . Don't . . . Understand . . . Then . . . You will . . . '

HITDOYF1, 'To die . . . now.'

HITDOYF2, 'I . . . Now.'

The Secret . . . , 'Consider . . . Now Here.'

Evolution To Self-Consciousness, 'Oh . . . We are what . . . A Random . . . Ah . . . '

See The Black Dungeon Doorway—3rd Edition—Published by AuthorHouse UK: 'The globe . . . A long sparkling . . . It appears . . . Pyre . . . With . . . shoulders.'

'What is . . . The white . . . If that is . . . A glowing . . . companions.'

'It's time . . . The rest has . . . see.'

See Apotheum Colluseum—The Ultimate InterActive™ Game—2nd Edition: POP = Known—Unknown, 'Spiralling is . . . Plane.'

Power Trip, 'If you feel . . . From the GAP's . . . Plane . . . Being a God and/or Goddess . . .'

3. The Circle

A. The Circle is the 2nd Form of Motion.

The Circle is the Great Form of Life.

It is also the prerequisite to Self-Consciousness. Until the Power and Energy loops back on itself it is not familiar with itself. Forming the Circle are the Element's of Earth, Water, Air, Fire and Ether can be seen as going through it. They provide the closed circuit flow of Particles, Lines Of Power and Lines Of Energy in such order. See Apotheum Colluseum—The Ultimate InterActive™ Game—2nd Edition for more Element's.

At least two opposite Element's are needed to provide a closed circuit flow. The Circle is also the second most powerful and energetic Form of Motion.

B. Meta-Science: Observation and Experience: The Circle is based on the obvious workings of Nature. All things cycle i.e. the Season's.

C. Science: Mathematic's: Geometry: The Circle contains all the regular polygons within it.

Physics: Electricity: A closed circuit is mandatory for Electricity to flow. Otherwise it simply disperses. Electromagnetism: Positive and Negative poles are need for flow of Particles, Lines Of Power and Lines Of Energy.

D. Philosophy: Theosophy, Eastern Religion's and Philosophy's and many others: O, The Circle is a universally accepted symbol for Eternity, of the equivalence of Beginning and End, of Life. The Snake/Serpent is biting its tail. There is also: ⊙ which also stands for Infinity with its 'center at nowhere and its circumference everywhere'. There is also: ∞ which are two Circles combining into two Spiral's in two linked Sine Waves for 'the shape of the Universe is an hourglass' and 'Life is a Sine Wave' and 'Heaven Planes are not the Hell Planes' and 'Life is a race track'. According to Jainism you simply have to make it vertical to see it better and then combine it across all possible Infinite Planes Of Macrocosm and Microcosm which is also supported by Hinduism, Buddhism, Taoism, Confucianism, Ayurvedic and Chinese Acupuncture and Accupressure.

D. Literature: See Poetry Lore—1ˢᵗ Edition—Mass Energy, The Death-Life Conspiracy—Published by AuthorHouse UK: Circular Energy, each poem.

See Poetry Lore—1ˢᵗ Edition—The Power Of Release—Published by AuthorHouse UK: Thank You Mom, 'This . . . to.'

Power Trip, 'I trust . . . Purgatory . . . Next to . . . Now it's . . . Release . . . We . . . want.'

All Things, 'All things . . . Work in . . . people.'

On The Same Street, 'Isolated . . . No binding . . . Only . . . If . . . It's. Safely.'

Progression, 'Each Plane . . . The way . . . Where . . . From . . . Circle . . . Plane.'

Collide-A-Scope-Worlds, 'Come . . . Everyone . . . It's fun . . . When . . . There is . . . Forever.'

'And . . . With our . . . So let's . . . And dance . . . body.'

Gates Of Dawn, 'Here we . . . We're going . . . Like in . . . And now . . . Dawn.'

See Poetry Lore—1ˢᵗ Edition—Self-Consciousness, Law Of Unification (all bodies)—Published by AuthorHouse UK: Self-Consciousness, 'My guiding . . . What to. Allow . . . With . . . To make. Poet!'

Friendship, 'Journeying . . . To hold . . . Having . . . Love.'

You Are A, 'Hello . . . Shout . . . Bring it . . . Z.'

Cause And Effect, 'Life is . . . As you . . . The only . . . Do not . . . Millions . . . Light.'

Self-Efficiency, 'Yea . . . Don't Understand . . . Then . . . You will . . . you.'

The Secret . . . , 'Consider . . . Now Here.'

Truth Is Commonality, 'All things . . . Earth . . . Like . . . On a . . . In . . . Eyes.'

Evolution To Self-Consciousness, 'Oh . . . Could it be . . . From . . . Of the . . . Circling . . . Justice . . .'

See The Black Dungeon Doorway—3rd Edition—Published by AuthorHouse UK : Evident throughout system

'Then we stopped . . . damned.'

'We walk around . . . The demon . . . Me neither . . . I will remain . . . consciousness.'

'It's time . . . Everything works . . . What is life? . . . The prime . . .'

5. The Sphere

A. The Sphere is the 3rd Form of Motion.

It has the most Power and Energy. It is The One.

The Sphere contains all the other Form's of Motion as shown in the diagram of Poetry Lore—1st Edition—Self-Consciousness, Law Of Unification (all bodies)—Published by AuthorHouse UK. The diagram explains how it works.

It is the One Form which can embrace Everything, or is Everything, the Universe. If you combine all the possible Multi-Universes, it can also embrace all of these . . . The Perfect Sphere is the Form of Perfection, of GOD. The Perfect Sphere is the Form of Omnipotence And Omniscience, thus Infinite Power And Energy.

Nothing can destroy this Shape. It is the perfect relation of ALL Element's. ALL Particles, Lines Of Power and Lines Of Energy 'spheres' through this motion. See previous chapter here on Real and/or Virtual State Vector's.

B. Meta-Science: Observation and Experience: The Form of the Sphere is evident in the shape of our Mother Earth and other Celestial Body's, their atmospheres, their EM Field's, plus their Orbit's. After all, the Ellipse and Egg is the little sister of the Circle and Sphere, or big one in their cases. ALL Life is contained within the Sphere, whether it be on our Planet or another.

C. Science: Mathematic's: Geometry: All shapes can be contained within the Sphere with allowance for a fluctuating radius. The Sphere is the most symmetrical, balanced and Perfect Form of all Shapes.

Physics: Electro-Magnetism: The shape of all EM Field's resembles a Sphere. And all things possess an EM Field.

Astronomy: Astro-Physic's: ALL Celestial Body's have an Invisible to the naked eye spherical Spectrum of Energy around them, not excluding Galaxy's which in a photo look like a sunny side up cooked egg, blatant Anti Type-0-Negatives obviously forgot to stick on their Gamma Ray Glasses.

D. Philosophy: Theosophy, Greek and Hinduism: The Celestial Body's which are also Spheres were and are associated with God's and Goddesses i.e. Mercury is God of Literature And Trade while Aries is God of Battle And War while Luna is Goddess of Emotions And Empathy while Maya is Goddess of Illusion And

Perception. The God's and Goddesses are Immortal. Though not necessarily Perfect, once again 'Does one need to know the whole Tree or just one Branch to become Enlightened And Immortal?', they are the closest to perfection of all Living Being's, except GOD. Still though, once again 'What or who is GOD without his Henchmen?' A Moot Point is also still not a retort.

Many Religion's and Philosophy's: Though it may be hinted or cloaked in deep symbology in many Philosophy's and Religion's, usually in association with Divine Being's i.e 'radiating light encompassed . . . ' it is Meta-Science which states here the exact Nature of the Sphere, as well as Perfection, and it is not an Independence Day Conclusion for Meta-Science is for all People of the World's drawing from many sources but primarily: Observation and Experience (so lacking in many beliefs), Science (without evidence and proof there is nothing), Philosophy (Logic, Reason and Wisdom), Literature (there are many great Writer's and Poet's throughout History of Humanity who can now be found on Internet), Art (also now many on Internet), Music (some say now 100+ million tracks on Internet as everyone wants to make there own song) and Religion (I still think this is a Right Of Choice in your own Heart not excluding multiple choices).

Unique to Meta-Science and its system is the correct definition and usage of Observation and Experience, one can never deny the faculties of Human for new and Individual thought based in perception of Reality's around us . . . it is possible to become Enlightened And Immortal if you also apply Objective Observation with your own Subjective Experience and vice versa otherwise one ends up in to Heavy Individual Or Group Bias which most systems suffer from.

D. Literature: See Poetry Lore—1st Edition—The Power Of Release—Published by AuthorHouse UK: All Things, 'All . . . Balance . . . Are Based . . . This is a . . . people.'

Knight In Shining Armour, 'The Knight . . . Which has . . . Can he . . . And . . . Avalon.' 'And if you . . . With our . . . With my . . . So let's . . . You . . . body.'

Coming To . . ., 'Beautiful . . . Goodnight . . .'

See Poetry Lore—1st Edition—Self-Consciousness, Law Of Unification (all bodies)—Published by AuthorHouse UK: Front Cover, Diagram of the Sphere.

Self-Efficiency, 'Yea . . . Don't . . . Understand . . . Then . . . You will . . . you.'

True Love Is Made, 'The Sun . . . Made.'

The Secret To Happiness Is Now Here, 'Consider . . . Now Here.'

Truth Is Commonality, 'All things . . . The Four . . . Holding within . . . In . . . Eyes.'

Evolution To Self-Consciousness, 'Oh . . . Human.'

Quote: Golden Verses Pythagorean's: 'Pay . . . By . . . more.'

Society, 'Act I . . . end.'

You Have To Fight For Freedom, 'I . . . It does . . . Gaining . . . To the Sun . . . Universe.'

Freedom, 'The rocks . . . through . . . around . . . to an edge . . . beyond . . . wings.' See Poetry Lore—1st Edition—Self-Consciousness, Law Of Unification (all bodies)—Published by AuthorHouse UK: Back Cover, Diagram of the Sphere. See The Black Dungeon Doorway—3rd Edition—Published by AuthorHouse UK: 'Cool. Beautiful . . . We wake up . . . vanishes.'
'Orthe, Wodora, Aera, Pyre . . . forever.'

Like I stated in a previous chapter here this is also a cross-reference to my other works next to many sources on Internet. See also The Free Show v. 4.4, a lot of people made the mistake of not scrolling down far enough and still have v. 3+ which the GURPGS and my two websites Silverlingo.com and Planesofexistence.eu for other sources.

Free Democracy

Whereas this idea is not new, it does warrant discussion: Free Democracy is the form of democratic Government of the future; the point is, of course, to apply it on a global scale as part of a Global Economy and a Global People. As a Canadian and Netherlander, I can ascertain the need for this! I also am a Global Canadian Citizen which everyone in the beginning of 21st Century seems to have completely forgotten.

The concept of Free Democracy encompasses: Free Voting, Free Choice, Free Speech, Free Thought, Free Information, Free Economy And Politic's, Free Religion, Free Science And Technology, Free People, and most importantly, Free Welfare. Unfortunately, also these days, I seem to have to on a daily basis retort their blatant finger pointing at Socialism which Free Democracy is definitely not, so once again go image and label yourself in your own mirror and rear view mirrors.

This essay will deal with each of these 10 Major Freedom's plus other Minor Freedom's which fall under multiple Major Freedom's.

0. Law

You have no right to take away any major or Minor Freedom from an Individual, Member and/or Group unless there are Witness(es), Evidence and/or Proof to be prosecuted by an Official Judicial System of a Country in Free Democracy as a Criminal, Terrorist and/or Enemy who has recinded his/her/their Freedom's in Free Democracy.

1. Free Voting:

The most important factor of Free Democracy is the right to Free Voting; through regular Election's this often decides the fate of nations. Not only should this be without cost for paying for Free Voting would be absurd, despite the ridiculous amount of Money spent on Election Campaign's, but it should be greatly expanded upon. With Internet, despite all the hacking these days, it is technically possible to provide Free Voting on many Topic's Of Debate and to provide such much more frequently, not only once per 2—5 years for a Representative but by utilizing Pop or Poll Votes across different quantity and quality of Population Group's as Netherland's is presently in 2014 doing we can get a fairly accurate consensus of the Will Of The Peoples; unfortunately here we have a little bit too much of a Split Faction Effect where anyone who meets the prerequisites can start there own Political Party resulting in even a couple dozen of them. This obviously splits Votes up too much and so is very heavy Coalition based. Each Political Party has a Representative by the necessity of a middleman to primarily translate technical jargon into laymen's terms with the People. This representative with the vulnerabilities, weaknesses and corruption of such a Figurehead is supposed to continuously adapt and meet the Will Of The Peoples; here as you see I am leaving out the singular since it is now being very clearly shown in almost every Country on Planet Earth that there is hardly any Nationalism left; this part of my Country is not that part of our Country, this part of your Country is not that part of their Country and every other possible division.

Free Voting on Issues is done on a regular basis which is a variable depending on Degree Of Priority of Issue. This, of course, does not mean each and every day, such would result in Chaos And Anarchy. However, if each Community (a social group of any size whose members reside in a specific locality, share government, and often have a common cultural and historical heritage—Dictionary.com) of a Country do not have the right to Vote on Issues which affect them then the Will Of The People can never be represented correctly and we have no Type Of Democracy at all, just another Manager or CEO or President playing GOD. A board of only 3-10 directors or stockholders is also inadequate, leading to Power and Energy imbalance. Here as you see I can use the singular since it refers to specific Group's.

The old systems of Roman Republic, Feudalism, Western Democracy, Social Democracy and other attempts to simulate a Democracy (still Demo Beta Version 1.1322661831) across previous systems such as Dictatorship, Communism, Monarchy, Constitutional Monarchy, Republic and Capitalism by primarily representatives of a large faction range of Political Party's can only be seen as antiquated in Free Democracy. This is primarily due to the responsibility is too large for one shoulders, thus the attempts to form House Of Common's, World Congress and United Nation's. Free Democracy works toward in the future to more accurately provide for the Will Of The Peoples and not only the 01% Rich Elite by primarily proposing a better Voting System and the following 10 Freedom's of Free Democracy.

In the far future, to hear them jeering already, one could even Vote on the weather in an EM Domed City, however for now we do not lack a much stronger sense of Realism and how long it takes for development to actually take place. Two expressions come to mind: 'Everything on paper never works in practice' and another one from my own Poetry Lore—1st Edition—Published by AuthorHouse UK, 'It's so easy to destroy, so hard to create.'

2. Free Choice:

Though an actual Top 10 Priority System of these Freedom's in that order may not be realistic or possible, Free Choice is seen by many in the beginning of the 21st

Century as one of the most important Freedom's. Anyway, always leave something to the experts . . .

Without your Freedom to choose freely what you do each moment of the day, the system only results in Modern Slavery, or what some amusingly call 'Volunteer Work'. Volunteering is good, but not when you have to sell your land for '1 peso' again . . . Somewhat less amusing resulting in even bloodshed and violence are the protested extremes of Neo-Communism and Neo-Nazism. I would even argue that these are not the only two for there is also the highly unreachable not existing Absolute Middle; next to the fact the mean does not exist, almost everyone laughs at Wonder Drug's, Miracle Cures, National Wonder's and anything which hints of Pure Absolutism Effect's; this is also supported by many who agree there is a strong Graying Effect in our Timeline.

If I cannot choose to go left at the corner, straight through her middle or right at the intersection because someone is standing there with a Semi-Automatic Rifle then where does one go? So, Freedom of Movement as a Minor Freedom is hand-in-hand here; it's only a minor one since it falls under multiple Major Freedom's such as Free Choice, Free Economy and Free Religion.

Also, if I cannot, as a responsible Adult, or even these days, an Adolescent make a responsible choice or decision then such is what psychiatry, detention and/or imprisonment are for. So, just as each coin has two sides, the lack of Freedom is tied into your and my Free Choice. A criminal, after all, has broken his/her rights.

Free Choice is also connected into the other Freedoms. If I cannot choose between 7Up or Coca Cola then it's only a Monopoly.

3. Free Speech:

Free Speech does not mean you can say whatever you want and Insult everyone under the Sun of Allah either . . . This should be particularity learned by those who are Racist and/or Discriminatory. Regardless of how you attempt lame retorts like it's your opinion you are still not a Global Canadian Citizen even. One of the very first things we learn in pre-school even is the difference between Constructive Criticism and Bashing.

It, however, does mean you have no right to Control Censure Channel an Individual, Member and/or Group for having a different Opinion and/or Ideology unless they are in Violation of Law. Without Free Communication, which I like to call a Minor Freedom since it falls under all the misinterpretations and mistranslations in every sector though most agree these days that IT is somewhat over-complicated, there is already no Knight's Round Table, First Instances, Middle Instances, Last Instances, Debate, Mediation, Negotiation, Discussion or even just Conversation, why don't you just grab the gun, now, again . . .

Free Speech does not support Violence and/or Violent Protest, regardless of who shot first across all the centuries and millennia again it is by definition in Violation of Peaceful Protest which is the right of the Will Of The People, though unfortunately not in many antiquated systems to date.

Verbosity is a far milder form, we are allowed to talk with each other, and it is still better some Japanese or a CEO throws a chair instead of a Bomb at someone's skull . . .

Without any of these Form's of Free Speech no signed deal can ever be made. This error in History Of Humanity to date is just mind blowing, how many times have you heard yourself some Verbal Agreement or Handshake Deal made? That is not binding and has lead to everyone's 131ˢᵗ Degree Of Black Belt Bullshit Artistry. Thus, without a Signed Contract like the Geneva Convention Accord signed by Majority Rules of Country's in Free Democracy then it is not binding. Hopefully, of course, in the near future we will all be a little bit more Near-Enlightened and not fall, once again, into one of the Top 10 Major Error's in Democracy and Capitalism called the 51:49 Minor Victory. We have seen this happen so many times now if we don't adopt a better ⅔—⅓ or ¾—¼ Major or Landslide Victory Vote Result then the problems will just keep recurring as everyone, especially in UK, CA and USA Politic's, go and block each other again which also results in no actual solutions to real problems.

Without an Opinion nothing can start.

Without an Ideology nothing is ever reached.

'The greatest Innovation's and Invention's started with a new Idea . . .' is still and always a great quote.

Even better is, 'Ideas and Ideology's have always decided and determined the course our Timelines . . .' to paraphrase a little bit.

So, don't let anyone shut you up either, otherwise with various Trigger Word's and Trigger Phrases you could also trigger The Shut Up Terrorist.

4. Free Thought:

The right to Copyright and Intellectual Property (from 1915 CE) remains the strongest Law made by Humanity. Even though a lot of things these days on Internet primarily are being ripped off through Loop's, Weaknesses and Holes in the Law's of your present Not Free Democracy Country through primarily Minor Unacceptable Variation's except by Local Reduction such as a slightly different name and/or land extension it still is First Come First Serve and if you also happen to have carbon dating like I do for these Evolutionary Essays, Spy Kills, Spy Kill's 02 and practically all of my other works going back to when I was a kid even when I wrote my first Science Fiction Fantasy Horror Short Story and the first part of All The Things I Like poem and as a teenager from 14 years old The Black Dungeon Doorway (inspired by AD&D 2ⁿᵈ Edition) and from 21 years old Evolutionary Essays at York University in

Toronto, Canada then there is very little dispute by any Judge and/or Jury as to your Ownership. I always recommend as in my different Types Of Open Source which I call FREE Draft's, Final Editt's (purposeful misspelling) or First Run Of Final Edit as is the case with these essays that you place your Real Name and Copyright on each and every page; as a personal style choice I'm just not into Pen Names even if they do authenticate across your Real Identity. Nicknames are fun but do not constitute your Real Identity unless it's also your Artist's Name, Icon and/or Trademark i.e by using these as an example, Tom Cruise, Jim Carey, Kate Beckinsale or Jodie Foster, am I technically not in violation of such or is there no problem with a reference, fragment and/or quote again . . .

Before then Dictator's, Emperor's and Nobility decided everything, not to mention their Church with all of their Lines Of Power and Energy. See previous chapter(s) here. Once again, a perfect example which can also be found at Wikipedia is the following: Is it Bible, Holy Bible, The Holy Bible and/or The Bible? Thus, equally, and very important for Unique Definition's is it The Church, Church, Holy Church, The Holy Church or can they be only called Vatican and/or The Vatican? Like I stated in a previous chapter I did not fail to utilize this with The Open Markets. In my system I turn it into The Open Market's since I use this Name Convention in my GURPGS which by the way, just to dispel old conflicts between them and Gary Gygax, Steve Jackson and others, I happen to be like the only system that has the most comprehensive list of Christian Religion's.

No one can ever take these words I write here and no one can ever prosecute me (presently only in Modern Western Civilisation) for it. It is also a Nihil Probability that even one paragraph only, almost even one sentence, some say even word combinations, are identical to anyone else in Existence.

'The only way to stop me is to shoot me, right in the back of the head in the soft spot, and if that's all you are doing is Insult's And Provocation's then you deserve it.' states Mr. Newbie, Rules Lawyer.

'Next thing you know, they'll take my thoughts away . . . . Polaris! Polaris!' is a great lyric by Megadeth, especially these days when we see everyone keeling over dead left, right, center and middle, but no one will ever know how I feel or what I am thinking at a certain moment. Only with Hyper Modern Technology and/or Telepathy and/or Magic and/or Divine Magic is it possible to actively scan someone's thoughts or emotions.

Such does also never take away my right to come up with a new (and much needed or wanted) perspective. Even Ideas, Opinion's and Ideology's which at first seem weird, strange, absurd, ridiculous, unlikely, dubious, bad and/or repulsive can be revolutionary; both Ford and the Wright Brothers were ignored and/or heavily criticized.

'They first try to ridicule an Idea to see if it or the bearer can be broken.'

Such close-mindedness equals fear and ignorance stemming in Humanity's instinctual distrust of Unknown due to too many dismemberments and mutilations in Dark Forest's. So, on the other hand, it's also a necessary Self-Defense Mechanism.

What Science alone holds for the Future, stigmatized by outdated laws, is Infinite . . . unless Black Holes swallow the whole Universe again . . .

5. Free Information:

You wold think this one comes right after Free Economy but without first thought there is no information, no action . . .

Our right to Open Source is necessary to avoid Monopoly's, even Group Monopoly's. This is also strengthened by the fact that many Open Source Project's and Product's are developed to Retail Project's and Product's just as is my own goal to turn all my works into. Not that I'm jealous like so many others of their SM Monopoly which they put all the Research And Development into since we cannot function without them either, not that I'm *only* an SM Lackey either cause I have an MCSE Diploma. See curriculum vitae.

'The Knowledge of Humanity is Free to ALL . . . ' is also a timeless quote, if you don't mind me applying my Name Convention to it.

Before Internet, Author's would suffer for years and decades before they got their word out, many even post-humously, even worse they practically never saw a single Red Cent themselves and then their works were sold for millions.

Contract's, Deal's and Agreement's made between Individual's, Member's and Group's is at the core of the nature of Humanity, but they also create lopsided Power and Energy Structures. The Manager or Editor, CEO or Final Editor, is playing GOD, already.

You and I have to be, globally, allowed to exchange Free Information with one another, without fear of persecution and/or execution and no longer suffer under their Control Censor Channel Complexes. We can finally prove this now too: Copyright and Intellectual Property is NOT Patent. As I have stated in my other works I will NEVER Patent anything in/of my theoretical propositional prototype thesis. Thus, technically speaking Evolutionary Essays is only one part of my whole thesis being a central pivotal point in the whole works. I don't want the Responsibility's and it is not my Department's; I also have a Bad Joke now: 'I don't need any more than a 115 IQ Level to destroy Planet Earth.' So, what are you waiting for, go out there and make yourself a millionaire.

'The root of all Evil on Planet Earth really is Money, since they stick it in only their own pockets, but we will also have to use Credit for Good too . . . '

Is a God Emperor supposed to rule the Planet Earth or do the people have no say . . . , 'All for one, and one for all, but is One really for All, and vice versa?'

'I mean, word up, word out and word around.'

'Knowledge is Power and Power leads to Corruption, Absolute Knowledge is Absolute Power and leads to Absolute Corruption, Absolute Knowledge is Absolute Enlightenment but one cannot become GOD.'

If it is not done by 01% Rich Elite and Will Of The People in the framework of Free Democracy in the future then the problems of antiquated systems will recur.

6. Free Economy And Politic's:

Anyone with $100,00 or €100,00 netto or bruto can turn it into $200,00 or €200,00 netto or bruto! Yet, not all succeed at selling their table or have the right to do this due to over-complicated Reseller and/or Patent Law's. The principle remains the same, however, if I have a table which I Own then I should have the right to sell it as long as I pay Tax. There are also Government Law's, especially here in Netherland's which prohibit someone with Types Of Welfare to earn Money, extra Money and/or included Money. The first and second fall under Fraud and the third is a regulation which allows you to work Part-Time and Social Services makes up the difference so you stay at a min of the normal level for people across Age And Status. On the one hand this provides better stability and fairness for people who are less able, young, old and/or handicapped but on the other hand when I can also not register a business in Kamer van Koophandel (KvK) due to outdated regulations then I prefer the Chinese who '. . . turn 1 cent into 2 cents into 4 cents into 8 cents into 16 cents into 32 cents into 64 cents and so forth and we in the West now make no sense and this now makes no sense at all . . . ' leaving me stuck in a Catch 22 across many instances here or a Viscious Cycle which I cannot exit.

The concept of Free Competition is therefore very important here, too. I have No Right's to earn Money on Welfare here. This is just not correct, with Internet alone I could even become a millionaire like many other already who started a webshop, however those outdated regulations as many other things trying to adapt, upgrade and/or update across Internet were made before Internet existed! Why am I not allowed to sell anything on Welfare?

Why am I not allowed to resell a table? Well, next to various instances such as Centraal Beheer Logistiek and also Centrale Boekhuis Logistiek which strangely enough have the same acronym one is not allowed to sell across Reseller's from AuthorHouse UK, Amazon, CBL and Bol.com, an Issue which took me 6 months to get through, ranking second next to the IRS Double Tax Issue (ITIN Form) between US, CA, UK, Commonwealth, EER, EC, EU and NL which took me from 2004 to 2014 to get through, a whole decade, when I published The Black Dungeon Doorway—1st Edition—Published by Trident Media, Inc—Washington House and never received any Royalty's and everyone including myself thought I was ripped off by Sam, my Agent. If you recognize the Publisher then you see they went on to make one of the most succesful Vampire—Werewolf Film Series ever called Twilight since Trident Media,

Inc. published (some) their books. Thus, quite literally, my books are only available at Internet Bookseller's like Amazon.com and not Bol.com since CBL is their own separate publishing instance. This is where the expression in our medium comes from: 'Now to plopagate to Reseller Galaxy . . .' Unfortunately, this extends across entire Planet Earth like 2 Giant's battling it out with 2 Big Club's and kills any sale potential I have here since the vast majority of people in Netherland's use iDeal directly with their bank account and not Credit Card's which are required for websites outside of Netherland's (EU Bank Issues included).

I would just like to appease everyone that I have no ill feelings or harsh regrets about Trident Media, Inc., Washington House and/or New World Media who offered to place their logo on my book instead of Washington House since before 2004 I was a practical nobody with relatively little Hit's and they helped me greatly in my Internet presence. In case you haven't figured it out yet, Silber, Psionic Warlock, the protagonist and Numero Uno Most Hit And Popular Character Class in The Free Show is both a Great Gray Werewolf and a Silver Silberian Tiger next to his other properties. After then in one year alone I even got +/—700,000 Hit's on my website Silverlingo. com and to date I now have at a min of 40,000,000 Hit's in/on Planet Earth. So, also, my thanks and gratitude, once again, cause your hardest step is to get published and presence for the first time.

The present systems of Money, Credit and Barter have many drawbacks, disadvantages, holes, loops and weaknesses; we try to give the blame to all of them as Highway Robber's but 'if your bare ass is stuck in the air then don't be surprized if someone gets tempted . . .' If an Economy, a Free Economy, is to succeed then these minus points have to be corrected. The worst to date, in my opinion, not that I'm an Economy Expert or anything, they sure as hell ain't either, is primarily: 1. Output and Input 2. Consumption and Resources 3. Money and Credit 4. Open Source and Retail Services, Project's and Product's 5. Corporate and Government Debt 6. Open Market's, The Open Market's, Free Market's, Stock Market's, Corporate Market's, Government Market's, Group Market's, Member Market's, Individual Market's and Black Market's.

'Half of EU is a Black Market. Half of Planet Earth is a Black Market.'

A far better proposal is a fully registered digital Credit System. This became only possible with the advent of Internet. Otherwise, your Economy will keep making massive losses with the Black Market, illegal copying and ripping, not per se the stereotypical blaming of Open Source. This, however, still won't stop anyone from making 10000 copies of a Retail Product and sending it to all his pals for only $1,00 via East Indian Trading Company Effect due to the nature of IT and XCOPY since MSDOS. It's not Bill Gates fault either for what would be the point of a computer which cannot copy? With registration, though, it at least gives your registered business the chance to trace the Criminal and provides Taxes to your Government.

Unfortunately, which I have addressed in my other works and forums on Internet, Internet to date has a serious technical Issue with Real Identity and First Instances across Roaming Profiles: You still want to logon remotely with multiple laptops. Also, your Real Identity is NOT your Nickname. To worsen the problem which if you ask us is actually the cause of all hacking, though there were always Criminal's, if you don't have a Unique IP Address with your Real Identity at your ISP then you can hack as much as you want and just Re-Image your OS. Thus, in other words, the Open, Wide and PPTP HTTP Broadband Issues do not fail to fuck everything up, especially PR and Sales. This part needs to, of course, be seriously analyzed by IT and Economic Expert's and Specialist's alike for according to our own PAT DAT the East has not failed to rip us off quite severely since the millennia it would make you go all Pale Face. Just one out of many real examples is the apparent ease which China and other Eastern Country's can Patent our own Copyright And Intellectual Property's before we go through our own much slower outdated processes. The next thing they will not fail in is catch up in the Space Race since we are suffering the worst Recession and Budget Crisis since Empire State Building jumping stunts.

As only an Individual one has no protection except for Copyright and Intellectual Property which remains strong, do not touch My Unique Combination, but it does not help my wallet . . ., 'Ya sure, I'm a Hit Generator, but I haven't earned a single Red Cent in my entire life, some say I've earned $200+ million dollars to date on Tax Residual's, not including Prototypes to come, but it's not my fault they blew it on High Risk Investment Plan's and Blow Real Estate . . . '

'Didn't I say buy the land and not the Real Estate, if you played The Telephone Game to that extent then you got what you deserved . . . '

On Internet we are presently arguing between primarily 2 Camp's: Open Source and Capital Hill. My point of view on this, amongst many others, is the difference in standard between quality. A Blu-Ray disc or DVD (5.1 to 7.1) is far higher quality, has a nice Retail Package (sometimes even with a small pamphlet) and is not low-grade copies of music tracks at 64 kbps—128 kbps which should be free. What other deciding factor is there? This is why I have FREE Draft's, Final Editt's (purposeful misspelling) and First Run Of Final Edit copies in circulation for otherwise that god awful 13th Monkey Effect dominates and I get even My Ideas ripped off due to Spying and Invasion Of Privacy. Another reason for this is simply, +/—90% of the population of Africa to date in 2008, and still to date in 2014, and the Average Income Family Unit can never afford a DVD collection, let alone Blu-Ray, it's Basic Math here in Northern EU Rip Off Zone and I doubt it's so much different in most Country's: 100 DVD's X €4,99 to €9,99 = €499,00 to €999,00 per year. Practically no one can afford this for only one per 3.65 days! I've said it more than once, no one can calculate the Bell Curve right, if you would each just drop your prices by 25%, if that's doable across middlemen, then you might be surprized by the results.

The concept of Bust and Boom is outdated with Internet which is continously growing: Bigger, Better, Faster is Stronger . . . I can also argue that a slow and steady gain Economy is preferable and now this one again which is almost sounding like AI or Insider's Information: It is now 2014 and Internet is about to tentuple in size since, it's a well known statistic, how many people only are so far on Internet out of the entire population of Planet Earth?

The point of Free Economy, despite the multitudinal problems which we face, it looks more like a horrific Hydra Hybrid Clone Complex with a Vertical Wall Learning Curve, is each and every Group, Member and Individual should have the right to sell something on Internet within this strong framework utilizing primarily Free Competition and hierarchal pyramidical cross-lateral departmental geometrical priority selling structures. This sounds a little bit like Networking but it is not for how else are you supposed to go about it: Is there no more CEO, Manager, PR and/or Sales? Technically speaking, if I really don't want to sleep anymore since Internet is 24/7 then I could do it by myself but I'll leave that one up to his other Pale Face.

It does not mean all things should be free. Free Demacracy, with a play on words which I like 'Free Domacracy' is about the development of more Freedom's in our Fight For Freedom for Democracy's, since, once again, according to MySQL Database 2008 there is no Democracy yet, we are in West primarily Federation's, Confederation's, Constitutional Monarchy's and Republic's.

7. Free Religion:

Each Culture, definitely from the perspective of a Canadian, has greatness in many fields. For many centuries, until the Renaissance and later Picasso who sold postcards, these were all intrinsically bound, just like all fields used be in Philosophy from Greek's. Then Specialization, Industrialization and Commercialism took over.

The belief in GOD, God's and Goddesses in Monotheism and/or Polytheism in diverse Pantheon's on Planet Earth is even more ancient.

Since the beginning of the dawn of Human, far from Humanity as of yet which required Napoleon, one has always asked oneself what the mysteries, origins and sacredness are in Reality, Nature and Life.

Even cavemen and cavewomen worshipped something as shown by cavedrawings.

Even Science does not deny the theoretical possibility of a GOD, God's or Goddesses, except for their more extreme 100% Atheist White Coat Empirical Group who not incorrectly to date state there is no witnesses, evidences and/or proofs for such.

Well, I still like this retort: 'Thank GOD for Science!'

As stated in a previous chapter here: GOD = One Big Reality and Infinite GOD = Infinite Reality. This does not fail through the whole hierarchy.

Regardless of what your perspectives, your beliefs, your convictions and/or your axioms are you still have the Right Of Choice Of Religion's. This means you even have the right within Free Religion to choose for multiple Religion's! Unfortunately, this to date does not apply to those Country's who practice only Monotheism though in beginning of 21st Century everyone is asking if there is really such thing as a Mono-Cultural Society left anywhere. Technically, due to such previous centuries and millennia they do not lack the right to impose such on their peoples but must there be so much bloodshed? If you interpret the precepts, tenets and paragraphs of Monotheism's slightly differently, thus the warring factions, then does it really say you can kill yourselves, your own blood or does it state you can kill the Enemy? The only thing not unclear to us in the West to date due to cross-cultural and cross-language barriers is you're only allowed to conquer and/or convert if they do not have their own religious texts, scripts and books.

Oh my god, imagine such, I now have 11 Religion's, each in a variant fraction as a measure of my Strength Of Faith . . . This admittedly could get a little over-complicated and it is recommended to reduce your Choice Of Religion's to at a max of 8 Religion's, however who does not have all their Faction's everywhere these days?

Christianity alone has 24+ Faction's now.

And we do not lack our Celtic Christianity now.

8. Free Science And Technology:

Many similar arguments stated so far in this chapter also apply to Free Technology.

I and you without paying for overly expensive and over-complicated Right's Of Patent should be able to place a blueprint on a website of a futuristic Technological Device, like a 3D digital layered surround sound smartphone which you can aim at someone's head or buttocks from 2-200 meters away in Public and a hi-pitched Not Unhorny Blip sound without it being ripped off or wrecked by some Spy or Saboteur; once again, don't mind my Sense Of Humor since this is obviously a completely Illegal Technological Device Alert inspired and propagated by some Old School Perv who likes to finger his pocket.

This can be done live on Internet in 2 minutes flat. The date of the upload or the live edit, also the date on the paper it was made on, which is provable, and/or the date of scan is what counts. And not some long drawn out expensive Patent Procedure . . . Copyright and Intellectual Property and not the patent is what should decide it for the Technological Device Engineer. You can also stick a long random string which I like to call a Manual Encryption String so it's practically impossible to guess. Unfortunately though, Hacker's are more than a little bit clever and oh no I've been Compromized By A Noobie again which is just the worst possible combination, it probably wouldn't last an hour before they try to hack your whole website again just to Get The Info Alert. However, do you not have the actual Ownership of the 2D/3D

Object? This is also a highly debatable point which various Specialist's and Expert's need to target as to its viability and/or validity.

Otherwise, we get, once again, Power and Energy hungry Corporation's, Government's, Member's and Individual's like in Aliens and other Science Fiction Film's who literally kill their competition; well as far as I'm concerned with not irregular Hack Attack's on my computer systems hacking is no different, this is as they the quo pro status sum blo of how you even get a job just like with Facebook Employees. I can easily argue Counter-Hacker but have to admit again there are only 2 Types Of Hacker's, a big 'H' Hacker and a small 'h' hacker who do not fulfill the same functions.

If I happen to have an Innovation and/or Invention, and who can afford even 1 patent these days, plus the time it takes to be processed, then I should have the right to it, with proof of Copyright alone, not even Intellectual Property, to Patent anything I want by simply uploading it to my website in protected folder.

Given the Supply and Demand, I can then sell it or I pass it on to a Family Member through Inheritance since the Key Problem here is that I obviously don't want to make it and put it on the market right away for any reason.

In fact, the present State of Affair's with another Wrench And Bottleneck Effect in the Big Government Machine impeeds development of Evolution of Humanity i.e. to the stars.

A perfect example of how Multinational Corporation's also impeed this development of Evolution of Humanity are Sources Of Energy. I can also argue I want to make in my garage, with a landing platform, a Prototype Fission Fusion Flying Car since I'm not a camper like so many. If I stick these blueprints on Paper and/or Internet first then do I not have the Copyright, Intellectual Property and Patent to it and all its Income, Profit's, Royalty's, Payment's, Responsibility's and Risk's ensuing from such?

A Member or Individual last I checked with enough Money alone, though Power and Energy helps a lot, can pull off any stunt they want (and/or need).

If the Law does not protect Copyright and Intellectual Property across Patent's then it is already End Of The World Scenario which is, wow, what a coincidence happening right in front of our noses now on Internet.

Is it fair if one lives closer to the Patent Office or can afford it then one gets the Innovation and/or Invention?

Information takes 2 seconds to travel with microwave wireless these days . . .

9. Free People:

A serious black sore on History Of Humanity is it suppression, repression and brutalization of other Races of Human Species. We keep forgetting that we are the same Species and it amount to killing your own Brother or Sister.

'It's about the Species and not the Race, it's about Milleniaism and not Centurionism.'

And Humanity with i.e. AIDS, this much STD's and so many SAD's should worry very seriously about its survival into the future or suffer another Mass Extinction Event.

One can, of course, also argue not to repress other so-called 'weaker' Species, but it is not likely Humanity will ever stop eating meat; it might not even be possible, genetically, considering the genealogies. Here one must distinguish between Wild and Domestic Animal's as a Resource which is also very camp heavy both sides almost hating each other's guts again, but the fact remains we have always eaten meat.

It should be cut and clear, black and white, since the 60-70's when it comes to People or Race Issues: We are all just different shades of gray, for all those who are color blind . . .

You have no right to spit on a 'Black Nigga' or snub a 'Yellow Goon' or bash a 'White Trash' unless they really are in Violation of you as a Criminal proven by an Official Court Case, which as we've seen in 'Hick Land' or 'Nigger Ghetto' is a heavily slanted football or soccer field with corruption rampant on both sides going back to the Civil War with North vs South and West vs East, well, how much has really changed? In fact, such descriptives based on colors of skin are discriminatory, false and biased, automatically. It's more like different shades of 'Brown' counts for everyone if you get a Light Tan; if pigment decides your Freedom then we only descend into Barbarism and/or Despotism.

This we know as a Global Canadian Citizen, however their conflicts go back millennia and not just centuries thus it's a problem which could very likely never be solved . . .

'The Future is Brown.'

10. Free Welfare:

Finally, we get to the most important Freedom of them all: Free Welfare. Once again, I have to Kill The Notion and/or Squelch The Rumor that I am any Type Of Socialist anymore. I'll even be honest and say, sure, only once before did I Vote on SP in NL but since then I've only Voted on PvDA which unfortunately due to all the Label And Image Effect's going around and their present Alliance with GroenLinks does not help the problem; when PvDA is primarily a Labor Party even it's now falsely associated with the Leftist of SP themselves which is not the same at all in NL. I have also learned, developed, done Research And Development, changed, adapted and like my own Facebook profile I have clearly stated that I am now a Middle Middle Middle. To poke some fun at myself which so few do enough, I could also say Oppurtunist, Robin Hood, Aggressive, Independence Day and other somewhat more rude things though also in defense of myself, who is not these things on Internet these days, or anywhere?

This is meant also in Vertical Hierarchy and not only Horizontal Hierarchy, do not underestimate and/or overestimate yourself too much, know your Own Lines And Limit's.

My primariy argument in this matter which is backed up by thousands of examples to date with Ex-Military Veteran's, Homeless, Poor, Starving, Sick And Diseased, Handicapped, Youth and Elderly is that you and I would have already been dead 10 times over without Social Insurances. Due to the word and name again 'Social' which has bad and negative associations and connotations with 'Communism', 'Lefty', 'Leftist' and other swearwords as they like to call them since WWI, WWII, Nam, Bay Of Pig's with Kennedy's, the Russian Conflict's and Bi-Polar Opposite Enemy Syndromes of Left vs. Right going back to probably even the first triad conflict in China c. 6000 BCE and Mongolia, Korea and Japan, we might be better off using even Free Insurances which I can only scoff at myself since what bloody Insurance is now Free? As I stated in a previous chapter here this confusion in terminology is once again a serious Cause Of Conflict which will never be solved in the short term. Freedom Insurances? Now it's sounding C—Horror Film again. Demo Insurance? Hack Insurance? Do they even exist? 'Social System' originally meant via The French Referendum after France Revolution to use the same root word as 'Society' not per se Left with Communism, was not Napoleon an Enemy of Russia? Whether he failed or not in another erronous Winter Attack is irrevelant to the point here. I don't even think I know in this case what a better terminology is if that is even possible since it is now registered everywhere on Absolute Photographic Memory on Internet and even more importantly in all the Near-Photographic Memor of the People.

These Handicap Group's are not capable of working Full-Time, some barely Part-Time and most hardly at all. They each have primarily a Need Priority, not a Want Priority, and our present systems provide inadequate Re-Classification, Re-Training and Re-Integration to be a contributing Member of Society.

Some are even dysfunctional and in the worst cases of psychological patients, completely misfunctional or not functional. Other Criminal Psychopath's do not need to be argued to much having Life Sentences already but some could be treated since we are, at least in UK, Europe and some other Country's, now in Young Teenager Shoes of Medical Science and Neurological Biology which is no longer than Middle Ages across possessed-by-demon complexes or the Classical Freud and Jung Complexes where you could let it all out while lying on your back on a brown leather sofa. Now it is all about your Brain Chemistry, neurons, synapses, neural pathways and what is defect there.

Only a Fascist or Communist would kill such 'Weaker Member's' of Society who are technically necessary for development of Evolution of Humanity through bio-diversity. Any form of Totalitarianism usually wipes out the 'weak and useless' who do not serve the Mother—or Fatherland and in some cases are perceived as a threat even.

This is atrocious. Human Right's Organisation's are left with a near hopeless task until Corporation's, Government's, Member's and Individual's step in to prevent death and disaster by utilizing, applying and adopting the Freedom's of Free Democracy.

'If I have only 1 leg then do I deserve to become Homeless just because of the War I became Alchoholic and can not spell or done good grammar anymore?'

'If I have damage nervies due to cigs which you can get like everywhere for $3,00 or €3,00 then do I deserve to Get Fired! with no Social Insurance(s) or Pension's or Lasersuits me just fine? What do they put in the cigarettes, anyways, these days??'

Well, we do not lack a website here in NL now with a list of all the bio-chemical ingredients per Brandname of cigarette and sjek: Tabaksinfo.nl of the Government.

Do I deserve to get blamed for my 'weak' Genetic's?

Do I have No Right's in a 1 month trial period at a new job?

'The strong have always carried the weak and wounded . . . ' is what one thinks of here.

If there is in the future no better Social Insurances, such as get rid of the millions of Homeless, especially in Winter, then our Planet Earth is doomed to Hell, by lack of forgivance alone . . .

While 01% Rich Elite have the wealth of the world and the other 99% share low to middle income and Welfares And Poverty reigning across Planet Earth in the billions, not even only hundreds of millions, then there is no other alternative but to develop to a better Free Welfare System.

Conclusion:

In conclusion, Free Democracy does not mean per se Free Money, Free Homes, Free Heat, Free Electricity, Free Food and/or Free Entertainment, unless Rich Philanthropist's want to pull out of their Private Wallet's, it means our Right Of Freedom's which are presently badly lacking in all systems, except for some Middle and only Upper Middle, Upper Upper and 01% Rich Elite Classes.

Technically speaking in a lot of CA, UK, FR, NL and DE Welfare System's now they are applying a so-called 'Participation Law' or other variations which state you get to Keep Your Welfare By Work through Free Job's and Free Work, as they free-for-free, which are Volunteer and/or Low Salary Job's and Work leading, hopefully, to Paid Job's and Work in the future. I've done probably even dozens of these already and is how, since my own last Paid Work was in 2003 CE even, I have not failed to reach my Tier Of Development in a number of Specialty's and Sector's.

These Freedom's of Free Democracy are not privileges which only the strong Man or Woman who works Full-Time deserves, a lot are even Part-Time, they are for every Human, who is not a Criminal, who has not broken his/her Right's, as described in a previous paragraph in the chapter.

Especially, different Races and Women who have suffered the worst Inequality's in the past from these lack of Freedom's of Free Democracy. It is not saying that each Human is equal which is now considered an error in definition throughout History Of Humanity but that each does not lack their Right Of Function, thus per Specialty, Sector and Module in Free Democracy in your Country.

'Do not turn a Baker into a Butcher or you'll only get bloodpies . . . '

I hope to be there in the future, and it is no Utopia, even if it means in a next lifetime, though to explore the whole Galaxy is very tempting.

Peace to the World. Peace to All. Peace to You. Peace to Me.

However, realistically, we can now in 2014 CE and onwards only:

FIGHT FOR FREEDOM!!

Energy's

The Five Elements

Earth, Water, Air and Fire plus Ether: The basic difference in these four (plus Ether) substances is their density.

When I talk of Being's in Elemental terms, I mean the construction and composition of Beings in their various Body's, Form's and States.

The chart of chemical 'elements' is based on the chemical bondage of various numbers of Atom's, Electron's, Proton's and Neutron's together. Chemist's recognize the four basic States of Matter = Energy Field's + Particles: Solid's, Liquid's, Gas plus Ether. I consider Ether to be advanced Fire (possibly double or quadruple Fire) due to its remarkable action when exposed to Gas. Space is Nothing with only Light filling it, Ether fills it to a large extent, too, 'We now see thanks to Maya Spy Sat, again, that that is probably like wow what is between each Galaxy.' I mean this is even 2-Bit Obvious to a Dumb Stupid Platinum Blonde. Once again, don't mind my Sense Of Humor and I can also argue it with the serious official Zero Point Field Theory that 'there is no empty space *in* the Universe', 'in' being the Key Word for we don't know what is outside of the Universe, and 'Nature abhors a vacuum'. This could also explain the confusion in Eastern Religion's between Nothing, Nothingness and Ether; Nothing can only not exist.

All Being's are composed of a relation of the Five Element's. A perfect balanced Being allows for the free flow of the Soul, Spirit, Mind and Body's i.e. I, myself and me, too due to various childhood positive and/or negative experiences never progressed beyond and 18-year old Emotional Body and it fluctuates either way by two years depending on how much any of you want to Irritate Me, Anger Me, Piss Me Off, Enrage Me and/or Battle Rage Mode Trigger Me and/or my Private Computer System's one more time! Like, just yesterday again as we now have to hammer shut Win7 too now on a daily basis to all their Hole Hacking across Wiki Leaky BG Shit everywhere I scanned and found 4 Critical Error's, three of which were read as Trojan's of some kind; they need to learn to stop attacking our systems so much or be classified as an Enemy Terrorist Hacker and not just a somewhat 2-Face Hobby Hacker. This would result in a perfectly spherical Being of Pure Energy; the physical Body of most Creatures has too many limitations. Earth, the most solidly bound Matter contains all other Element's in it. Nothingness of Space hits the Earth across a Rate Of Time and Water comes from it; Nothingness hits Water and Air comes from it; Nothingness hits

Air and Fire comes from it; Nothingness hits Fire and Ether comes from it. You also have Bi-Polar Opposite Element's: Fire hits Water and a lot of steam results; Air hits Earth and depending on such rates again it blocks or blows it the hell away. Above Ether could be Light itself. Above Light i.e. the Speed of Light can only be Nothingness, thus Ether, as was heavily debated by Einstein and those who opposed him. However, as I stated in a previous chapter, the Top 10 Element's are conclusively: 1. Earth 2. Water 3. Air 4. Fire 5. Ether 6. Shadow 7. Light 8. Form 9. Spirit 10. GOD . What is interesting thus is not just their number positions but how could Ether be above Light . . . This is what Einstein also argued and they proved it too.

Thus, Insta Travel and/or Near-Insta Travel through Space Time is theoretically possible; the latter is more realistic for how could anything even Spirit take no Time.

See Apotheum Colluseum, 2ⁿᵈ Edition for a more comprehensive system of how the Element's InterAct™ which is the Battle System of The Free Show, Spy Kills and Spy Kill's 02. This, indeed, also governs Relational Argumentation's, Battles and War's.

The cycle repeats. It is probably necessary to say here, I am not making this up. There are many reference in many systems to the Hierarchy of Element's, each with its own angle. Thus, it is also definitely Not Only Vertical, again, oh thank GOD, for once a cross-lateral system to balance out the whole Sphere, I mean how many times have we heard it from their Blow Suck Hurl Anti-Camp's to date . . . Honestly, I think they just like being Anti for Anti-Saki's Sakes, 'Thank you, Final Retort, click message left.'

These are the actual chemical reactions of the Element's when a vacuum, Nothingness, for they have now affirmed that there is actually plenty going on in such a so-called 'empty vacuum' which is now a misnomer to date, is applied to them; the 'vacuum' thus the Ether seperates the chemical bonds. In effect, you could say all the Element's (even chemical ones) come from the Earth when subject to this process; after all Matter is easily transformed into Energy these days . . . nifty, it writes itself not just Conqueror Rewrites History Effect which we have all badly suffered from for millennia long.

Through Evolution we fill the 'Gaps', the vacuums, which have Ether and sub-atomic quantum activity, until all substances are gained, resulting in Self-Consciousness, a Circle, when the energy loops back on itself and recognizes itself in a mirror, through development of Evolution of your Human it eventually grows into a more whole Sphere.

When we gain a Perfect Balance of Element's we gain Perfect Form, a Perfect Sphere.

Mother Earth is a good example. Of course, these Element's are found in Nature, in one degree or another, at one point or another throughout the entire system, not excluding Extra-Terrestrial Celestial Body's and in Space Time itself, debris is fun too. Mother Earth has primarily all five, just look around, though it obviously does not lack the next set of five. Consequently, all our needs and wants can be had from Mother Earth, if we would for once reintroduce Self-Sustainability for otherwise how did we survive to date through all the millennia; this is not green, they are just Bad Ecomonic's. Needs are our survival. Wants are our liberalizations, sucking up all Resources across explosion of over-population rates are all our demises.

Since the Five Element's can always be found in varying degrees in any Solar System, beyond Mother Earth, Solar System, Shadow System, Space Sector and Galaxy I would extend the Element's to all of Nature which is the entire Universe. Thus, the Element's are the common factor to all things and this is how all Being's evolve in Nature.

This to us is now unretortable; if you think you have anything but a lame useless retort to prove it wrong then I think you know where the postbox is rather than screaming your banal shit through the mediums.

Rocks And Stone

Primarily composed of Two Element's: Earth and Water.

The Eastern Mystic's believe stone possesses the first step in Consciousness throughout Evolution. They possess no senses, like touch, but crystals and precious metals are said to have a growth pattern. With the chemical elements found in Earth and Water these are the building blocks for all other Beings in existence. I could also argue you have to get a RDI of multi-vitamins and minerals in or your Hellth degrades badly.

Plant's

Primarily composed of Three Element's: Earth, Water and Air.

Note how much Fire, Ether and Light (the Sun's rays) Plant's need! Is this not Evolution of Species right in front of our eyes. These stimulate the 99% bio-chemical composition of Earth, Water and Air in Plant's. They are also all found on the ground, not in the air, like birds, cute little chirpy singing birds and not their screeching ones . . . Plant's also possess the sense of touch, as has been proven how they follow the course of the Sun.

As part of our preservation of Nature or rather as it should be put, preservation of our Species and other Species, and they have no conservation, we continue to grow and recoginze the value (especially in Medicine) of all Plant's in their role to providing all our Needs. Think of Alchemy, Hermetism, Organic Farming, Agriculture, Medicine and, of course, Biology and Chemistry. Without these, we are a lost 1-Man's ship in the middle of the ocean with a storm raging. Also a fact are the categories in Medical Science itself of Synthetic, Semi-Synthetic and Not-Synthetic which are all derived from Plant's.

This will lead us to eventual Utopia, though it could take a long time the principle still holds in Law Of Self-Sustainability. Trees and Plant's are the second most important, supplying 25-33% of Planet Earth's oxygen with phyto-plankton as Numero Uno Priority Alert across plastic oceans providing 75-66% and there really isn't so many other sources to replenish such across Animal, Human methane expulsions and Fossil Fuel Age pollutions.

Trees and Plant's also supply many other Need's and Want's. As our Heart's and Lung's are the air regulators and blood supply of the vast majority of all Living Being's, the Trees and Plant's are the air regulators and blood supply of Mother Earth. Here we also get good dosages of multi-vitamins and minerals from earth and stones which they eat.

Trees and Plant's purify the Air. Trees and Plant's hold the soil together. Trees and Plant's provide homes for many Living Being's. In fact Trees and Plant's with their Element combination provide almost everything to us so if we don't instate Law Of Self-Sustainability a.s.a.p. into ALL Sector's i.e. Hybrid's would already cut consumption of benzine in half then we ' . . . might as well hang ourselves by our own systems . . . '

Trees and Plant's hold the system in check: The major difference between this being a barren world and a thriving habitat are Trees and Plant's. Forest's and Jungles hold moisture in and prevent erosion, they provide a cooling effect from Ray's Of Sun.

Trees and Plant's also provide warmth and shelter in Winter with their bases melting snow and snow covered leaves and branches making little Igloos for Creatures in their midst, especially Insect's, Rodent's, Birds and also other Plant's for next Spring; Trees also provide shelter for other Plant's to grow, especially in a Jungle, 'Die by Hell Fires of Lucifer in die Amazon.' We're not going to stop repeating it, if that region goes, once again you could just Selection Harvest it rather than Burn Raze it to the ground while using only 2/3 of the pulp remaining and claiming it's fertile when it's Blood Soya, then you might as well look at Sandy Hurricane as a cute little baby pre-warning . . . Some even say that the Super Hurricanes were caused by such lack of moisture absorption and other imbalances, I still find it hard to swallow anymore that the three-in-a-row which hit Philippines, such being a huge shitload of water, could be caused by such unless you add up Amazon, Congo and Indonesia which were being strip harvested, such as according to Greenpeace being the size of Greece per year; it's more like some Dark Sorcerer is casting Mini-Black Hole Spell's and hurling it on his Enemy's heads, now who would there Enemy be we wonder, wow, what a coincidence, a Profile never lies.

If there are insufficient Trees and Plant's then other Plant's, Animal's and Human's will all die. At present Countdown To Destruction of Humanity according to thousands of statistics in all Medias we are on a Crash Test Dummie Hyper Acceleration Velocity Curve towards a Mass Extinction Event. The Battles and War's also triggered to fight for such dwindling Resources adds to For Whom The Bell Tolls Time Marches On.

So, when I say Trees and Plant's are the most important, I do not mean as *only* a resource. For each Tree cut down, two more should be planted . . . Trees and Plant's are the single most effective thing, next to controlling ourselves, to re-establishing Environmental balance through Law Of Self-Sustainability. As I stated in previous chapters and my other works we can also apply Animal Appartment's to this: Tree and Plant Appartment's cannot fail to succeed in Law Of Self-Sustainability.

Animal's

Primarily composed of Four Element's: Earth, Water, Air and Fire.

Animal's are all not stationary Beings.

Though many Human's exist in Animal and/or Poly-Animal States Of Being, a Human is different by Evolution possessing the higher Elemtents i.e. Fire primarily and now finally with dawn of Internet we are developing a lot more Ether. This is meant as sympathetic.

Since the Law's and By-Law's extend to all things and Animal's and/or Human's are living things, it is then conclusive to state they are our Brethren, possibly even our predecessors, and should not be harmed, but rather, protected, guided and nurtured. In a next life the tiger could become a Human; if you think this is symbolic, analogous, figurative and/or metaphorical then you have never heard of Chinese Astrology and/or met me, talked to me and/or listened to me since I happen to be a Tiger, and once again it really is NOYFB which Element's with all your Curiousity Killed The Cat syndromes.

Here, it is necessary, due to the consumption of meat from Animal Appartment's, from open fields and from hunting to distinguish between Wild Animal's and Domestic Animal's, and obviously the whole range in between. Everyone has made this error in this long-standing Debate between Vegan's, Vegetarian's, Omnivores, Carnivores and Survivalist's, which is not just B—Stupid Violent Black Humor, but quite literally Scavenger's which we'll all be forced to become if Planet Earth stays at this rate of consumption; another Bad Joke is we all in West only suffer from Consumption SAD's.

Animal's and Human's have the capacity for Pain and Pleasure; I was thinking of sticking an 'and/or' here but in my own experiences across many liberosities, Liberty Or Death Don't Tread On Me Symptom's, it seems each and every Pleasure for some inexplicable reason results in an equal Pain of some kind.

Animal's and Human's have Senses; despite Intuition having encountered so many Noob's to date I don't think the percentage is quite what you think. They exist on a primarily Emotional Mind and Body Level, the vast majority not an Intellectual and IQ Level one, due to their lack of Evolution to Fire, Ether and above . . . I state also Fire here since another great misnomer by practically everyone due to the limitation of the Visual Spectrum is there is Not Only One Type Of Fire; actually Intense Blue Silver White Fire is the most powerful and energetic of them all.

Thus, the Evolution of Animal's and Human's who are Intelligent, Conscious and/or Self-Conscious to different levels and degrees is to grow more harmonious, therefore less in imbalance caused by conflicts and consumption, with Element's of Nature for our Need's and Want's with Law Of Self-Sustainability through development of Evolution of History Of Humanity and its futures; we must not look at only the Past And Present which we mostly stare ourselves blind at but the Causal Focus Point Of The Future which will, to repeat one of my favorite expressions, either resolve the Timelines or they will simply converge and diverge again causing another Armaggeddon Scenario.

Human's

Primarily composed of First Five Element's: Earth, Water, Air, Fire and Ether. Secondarily composed of Second Five Element's: Shadow, Light, Form, Spirit and GOD.

As you know, only Human discovered Fire, most Animal's as with a lot of chips and organic ginger cookies, like a ducky I tried them out on, do a full 180° Turn About and waddle, jump, swim, run, sprint and fly as fast as they can away from it. If this isn't self-explaining then nothing is. Nicht Ontheil Lucifer!

Human's are the next generation in development through Evolution from the very first tadpole procreating to Apes to extinction of Ape Genealogy's to Homo Sapien throwing dung and hitting each other over the head with clubs, I prefer Dance Techno Club's, and possess all of the heightened quantities and qualities of Fire, Ether, Shadow, Light, Form, Spirit and GOD, the last one over the last 2.5 millennia sounding like an Understatement Nomination Candidate Of The Year Award; once again, this is meant as sympathetic by me, definitely not by them, for I perceive it as nothing more than the natural instinct, curiousity and intellect of Human to try and explore and figure out his surroundings, environment and eventually Higher Spheres Of Existence.

These range from Emotion's, Belief, Intellect, Reason, Logic, Consciousness and Self-Consciousness. Despite most Human's being dominated by their emotions and walking around in more like Jung's Archetypes Of The Unconsciousness and whole Computer Network's zombied and/or citadelled a Human can still grab a mirror and look at him and/or herself, even a little baby goes giggly wiggly. However, in my previous experiences with 3D Gamer's there was someone who brought up this interesting historical fact: There were no mirrors until which date . . . Is this an actual mechanism in development of Evolution of Humanity who could not look at themselves for millennia long or is it a bad narcistic Joke for you know who stared too long at his reflection in the water or that percentage had no Silver Jewelry or is it exactly why Water reflects Light . . . in any case it's still a perfect example to use here these days in beginning of 21st Century with all the blaming, shaming, laming, finger-pointing, in-fighting and factions dominating everyone, not just them though they are better at it now and zen, I mean, really, go and look in your own mirrors and rear-view mirrors and you shall not fail to find your own Root's Of Evil.

Human also possesses all the Tool's, Technological Item's, not just Magic or Divine Magic, as a result of Fire and Ether. I would also like to introduce here across Science Fiction Prototypes that it is theoretically not impossible to conduct the second set through such, in some cases literally like conductive precious metals, thus it should be doable in even the Near-Future to channel Shadow Energy, Light Energy, Form, Spirit and with the Will Of God launch it at the heads of the vile Enemy's who would once again dare and take our North Europe Stronghold's by many despicable means going all the way back to Khan of Mongolia and earlier centuries and millennia. It's only going to get worse too as their over-populations and environments are destabilizing, some say even America is once again greedily eyeing Canada for a massive Emigration Plan; if there is a red-stippled line between Canada and America then there is a thick red line between North Europe and South Europe. Don't forget it, again.

This is now self-evident to be the cause of our very strong and aggressive Resource And Territory Competition which despite all Bullshit Complexes on each side as to why they are all a Blood Cannibal to each other and despite all War And Peace Theory's to date is the Primary 0 Cause Of Conflict; like in IT I also use the 0,1,2,3 . . . set and not only the 1,2,3 . . . set for a 0 Object to us stands for 0 Core or 0 Root section of drive. In English and other Human Languages though if you don't place the word 'Primary' with it then like no one will have any clue what you're going on about.

'They are not per se Good, Neutral and/or Evil, they are not per se Stupid and/or Smart, they just have Relative IQ Level's.'

It is quite obvious these are in different degrees with different individuals. Look at your Western Horoscope too, to see how many Fire signs you have in your 12 Houses, and more importantly have a look where your own Aries is . . . The Numero Uno Best Family Joke of 2013 for us is still my Mother's Personality House, the 1st House, is in Aries and my whole Family except one all have at a min of one Aries somewhere, and my own Aries is in The Moon itself, not just Full Moon, so you better stop aiming shit at me punk cause I'm an Aries half the day, not only a Pisces.

We are, on Mother Earth, the most complex Living Being, the most evolved, despite the percentages we do not lack 01% Rich Elite and Elite Scientist's And Artist's. This may not seem like much in 20th Century Warfare, but 21st Century Warfare with Smart, Stupid, Space and/or Information Technology will provide hope or downfall . . . I would like to add one more thing here which is also a primary point in my other works, if we also don't bring 'weaker' genes for bio-diversity with us to populate a Colony Planet then we will be doomed to failure. This is the Primary 0 Error in History Of Humanity with its worst culmination at WWII, but then we are now getting International New's Medias Report's in that even in 1994 there were 800,000 Tutsis slaughtered by Hutus in Africa. Source: SCA, Aug 2013 CE.

Alien's

Primarily composed of ALL Element's: Earth, Water, Air, Fire, Ether, Shadow, Light, Form, Spirit and GOD.

Some here would argue that there is also SATAN as an 11th Element since it is still highly debatable as to whether there are ONLY 10 Element's but in my system, except for Nicht Ontheil Chronos! being the 12th Element of Time which is erroneously classified by everyone in Science, Quantum Science and Mathematic's as the 4th Element, I use Reduction and include Lucifer in Fire and Satan in Shadow. The problem with western thinking due to Classical Emperialism is that we perceive numbers, lists and systems as Only Vertical Hierarchy, the last 2.5 millennia of Monotheism domination has not helped this problem. You can also see this Top 10 Element's List as horizontal and/or circular. It's also Not Only Linear, you can draw hyperbole lines from each to another.

By logical extrapolation and where the Alien's are located (in Space Time) one can presume there are Alien's who possess higher quantities and qualities of ALL Element's. However, through Evolution of Galaxy's in Universes from Big Bang Theory or other Creation Theory's this has become a stereotype played all the way up to the hilt, not that they suck, quite the contrary, by Hollywood and is flawed. By such fact over Time there can and has to be also less evolved Alien Species.

Alien's possess Hyper Human Capability's and Hypo Human Capability's just like any other Living Being in different quantities and qualities. We can only speculate what these Capability's can be, even Paranormal Capability's such as Telepathy, Telekinesis, Clairvoyance and other more attuned Senses; Animal's have Specialty's in Sector's where they need better Senses. After all, Ether is not as inhibited in the transfer of Information and Energy's as other substances are . . . One of the favorite expressions of my Oom Han was that, ' . . . you can even run a computer on chocolate milk.'

Since these Matter's are still interchangeable with Energy's, it is not likely Alien's possess Immortality since they still fall under the Law's Of Physic's in the Universe. If they are Really Glowing Light Blue Spirit Alien's then they could possess Immortality.

A. Light Being's

As I also utilized in the 2nd Part of my trilogy, Planes Of Existence—1st Edition—Published by AuthorHouse UK, and Apotheum Colluseum, 2nd Edition which is the Battle System of The Free Show we can apply these classifications and categories in a comprehensive Valid System which allows for all Living Being's.

Primarily composed of 7 Element's: Air, Fire, Ether, Light, Form, Spirit and GOD.

As you see by the number Numerology going back to Oracle Of Delphi is not only New Age Mumbo Jumbo and plays a role in all things, especially your Calendar's.

Rather a God Being, a Light Being would definitely possess Immortality. This Type Of Light Being is not destroyable except by an equally powerful and energetic Shadow Being. It also having gone through Lower Element's and steps in Evolution is reaching a Near-Perfect Sphere and lives in Space Time; one the greatest errors made to date by overly Classical Emperialist's is they don't see the Spirit in the Sun's Body. Due to the relatively low level of Evolution of Human these Light Being's are probably cloaked and not visible to the human eye just like about everything to date in History Of Philosophy and History Of Science; in fact, the whole point of Science is to bring the invisible to the visible and the Unknown to the Known. Those who state or purport otherwise don't even know 1st Year Of Hyvescool or are Worst Not Unbiased Camper's. Light Being's also probably don't come anywhere near to Planet Earth for fear of causing a world-wide Panic Or Revolution which Humanity is not ready for . . . This is also the excepted Theory by the vast majority as to 'why have we not met Alien's yet?' They are A. Superior and don't want for us to Mass Panic Or Revolution effectively wiping out the entire Species or B. Inferior and are scared shitless of us. There is very little happy medium for the chances in all of Space Time in Galaxy's in Universes that there is an Alien Species with about the same Level Of Development In Evolution is about Near-Nihil. In any case, the very Act Of Interference would cause Chaos And Anarchy off the scale both ways.

Since there is very little scientific exploration into Light Being's and/or other Types Of Alien's, except by Hollywood and a couple other Sector's, it is difficult to speculate what quantities and qualities of Universal Element's they may or may not possess, but one of them is surey Enlightenment . . . See Buddhism.

They might also exist in Higher or Lower Dimension's such as 4th Dimension and above which is where the confusion is probably coming from: Is not Time the 4th Dimension and not Element? I put this in the present tense on purpose.

Dimension's

Dimension's are constructed Lines Of Force. They can also be called Lines Of Power and/or Lines Of Energy, which are coinciding, intersecting, parallelling and/or interacting all the time. A Line Of Force is a stream of bonded Matter with Particles and Waves through Time; like I said despite a strong Group stating it's one big EM Field Continuum I cannot see there being No Particles due to Monad Theory. See Theosophy and Higg's.

Following is first a List Of Dimension's and see diagram below named Crux Diagram:

0D: Thanks to Zero Point Field this is now Not Impossible. As to what or whatever is . . .

1D: The first number forms a 1D Point. Thus, I'll leave it up to you to speculate across the Being's above whatever the Hell is a 1-Dimensional Being but come to me is only, so very lonely abode, a B—Stupid Creature.

2D: The second number forms a corresponding 2D Polygon to the Dimension. Thus, apparently some Being's live in lower resolution realities . . . Jealousy Attempt.

3D: The third number forms a 3D Object which we live in, depending as to whether Time is 4th Dimension or not. Unfortunately, some people do not have that many = SAD.

4D: The fourth number forms 4D Virtual Volume.

Now here is the error made by everyone to date, is not Time bound to ALL Dimension's?

That's coming again across like Time = 0. Time Does Not Exist, like my Poetry Lore—1st Edition—Self-Consciousness, Law Of Unification (all bodies)—Published by AuthorHouse UK which is just that Great Art but otherwise I would not back it up to save my own life either. The meaning of that poem is we put our own Relative Measurement's on Time and claim to know what it is which is nonsense. As according, again, to Pre-Socractic's we can put an Infinite Quantity Of Point's on a Line, therefore to allow for Infinity it has to loop around and bite its own tail like the very well known Snake Circle.

Also if Time is supposed to be the 12th Dimension, Nicht Ontheil Chronos!, then it's coming across as Time bound to and going through Time itself which is over self-redundant and can only be invalidated.

Dimension's are no more than System Structures in Real or Virtual Space Time. Thus, it is still unresolved and highly debatable as to what Time relatively really is.

It is very difficult to construct a Form higher than 4D. If more Lines Of Force are added the Virtual Form simply becomes more dense. This shows Dimension's are merely a construct of other Dimension's which results in Reductum Ad Infinitum and technically we'll never be able to draw a 5D Form unless some God Computer shows up.

The 4D Form is not a Hedron. However, it is a legitimate Virtual Form as the 4th Line Of Force added is not redundant. It also actually looks virtual being not closed. It is not unstable due to Symmetry. Maybe the 5D is Asymmetrical being an Odd Number . . .

The Crux Diagram gives the proper system for constructing Dimension's.

'Higher' is greater quantity and quality of Lines Of Force. 'Lower' is less.

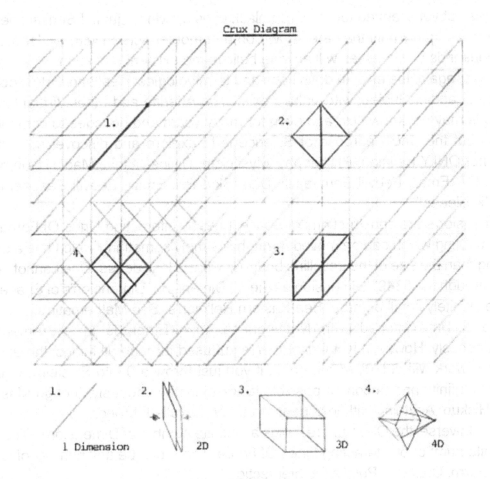

Crux Diagram

1. / 1 Dimension   2. 2D   3. 3D   4. 4D

Each Form 'con-forms' to the Dimension it is represented in.

This shows our perception of Dimension is relative to the point of the Observer and/ or Point Of Reference. This is necessary since in our Age Of Science, Technology and Information Technology some are trying to argue that a Hi-Res Photo and Maya Spy Sat does not lie about somethings . . . well, hmm, cough, do look her Goddess up . . .

This does not state Dimension's are per se Un-Real, as Lines Of Power and Lines Of Energy are very Real, it states they are Virtual with our perception of Dimension's as relative to the Point Of Observer. Thus, out of deference to their original terminology we could put the above two into this Object and/or Subject.

See how this does not even state perception is illusory, only deceptive. Our perception is not Un-Real, Virtual, it is simply relative. After all, I would hate to think I am not Real and only Virtual stuck in a Big iPOD Spaceship through Space Time to another Colony Planet cause we already destroyed Planet Earth and have relive all our Lucid Nightmare Reality's in a worst sense of a 3D Hell Matrix.

However, our Soul, Spirit and/or Mind's, not to mention Emotion's though violently expressed and visible, are Virtual. I posed this in a previous chapter here about overly quantitative positions in Quantum Theory's which still do not know where the qualitative Values are . . . are they spinning around our heads as smaller particles and lines not yet scanned for . . . deep black eyes glowing again, 'I Sensai she is not unhorny . . . ' Only until they are shown on a computer screen with evidence, proof and thousands of witnesses will anyone believe the existence of such.

This is, again, the error in definitions and terminologies: Real and Virtual do exist. An Object and/or Subject with Values exist. The Afterlife and other Virtual Reality's exist. We may not know now and even for many years and decades to come WHAT such is but the whole purpose of Science is to explore and discover such things, so it's not ONLY Religion, Philosophy, Mysticism, Divine Magic, Magic or oh oh, not again, 'Evil Email, Eeevill Eemaaailll, Dark Sky! Evil Email, Eeevill Eeeeeemaaaill, Dark Skiiiiieees!'

Dimension's are constructions of Space Time by interacting Lines Of Force.

Something which can be moved through is simply something which has a greater opening than the size of the travelling body. Or what is less dense. One cannot equally move through the 5349th Dimension as the 3rd Dimension. Dimension's of Space Time can be infinitely small or large. See Eastern Religion's. See Mathematic's.

Space Time infused with Nothingness, thus Ether, is in all Dimension's simultaneously. However, Not Only Ether is suffused through all things; the equation does not work without '0', smirk, get it, if you just throw a 0 into Einstein's equation then it is Infinity and becomes possible to travel instantaneously, though Mass is a Major Hickup. Anything with less density can be travelled through.

The Universe, the One Sphere, holds an infinite number of Dimension's. There are an infinite number of interacting Lines Of Force. There can be any number of Cruxes in one Form. Cruxes = Point's Of Intersection.

There is an underlying Infinite Timeline. There are Near-Infinite Timelines in a large Umbrella Form. There are Finite Timelines of Being's. However, this means by definition there could theoretically be Near-Infinite Being's and Infinite Being's. Extrapolation, interpolation and direct cross-relation are Not Invalid Method's.

Any Real or Virtual Volume can be reduced to one Line Of Energy, one axis being spun in all directions simultaneously. With a Sphere all the Point's Of Circumference are equally distant; this also counts for the Center Point.

No Line Of Energy is the Great Nothingness, Empty External Space, which is Infinity, there being just a Big Zero; how else can the Universe be expanding without causing the worst friction with what is out there and burn up the whole Universe again, like another god awful Coffee Cup Effect.

All Energy's, or Infinity, results in Zero. Infinity = 0. This is an Infinite Equation such as what we all learned: $2A + 3B + 4C = 0$.

However, there is also: $1A + 2B + 3C = 1$. The trouble with 3C is that you have to multiply 3 to .33333333333333333333333333 which goes on forever.

If we stick in some minus signs does it help: $4D + 5E - 6F = 0$ or equals 1.

'No, I think we're all the way up Shit Hell Fire Lucifer Burning Creek with Not Paddles and Know Canoes with Leaky Holes as Barbarian Bowmen shoot at us . . . '

Density

There can be any number of Cruxes in one Form.

A Line Of Energy is a stream of bonded Matter moving through Space Time. At Cruxes, Point's Of Intersection, bundles of Matter are made. These Cruxes are like Sun's, to whatever fraction of the Solar Sun. There is now also the conclusive existence of Black Holes of all shapes and sizes, 'With the size of the object alone at port or starbord, a huge Black Hole, you got to ask yourself how they missed it, it's more like each and every other day of Planet Earth depending on which side of the Sun we're at . . .'

These Cruxes are barriers which cannot be penetrated, except by Great Force, like an off-the-scale augmented Jedi. Cruxes are bound particles of imperfect Spheres and the Sun is a Perfect Sphere despite massive fluctuations and solar flares. Planet's are Near-Perfect Spheres, thus Elyptical Spheres, Orbit's are Elipses and we go around in circles all day long too . . . To penetrate the Cruxes themselves, not just moving through less dense EM Field's, requires destruction of their bondage.

Number of Lines Of Force and Cruxes in a Form is directly proportional to its Density.

Lines Of Force and Cruxes decide Density Of Dimension.

Density Of Dimension = Number of Lines Of Force + Cruxes.

Thereby, a Form equally or less dense than another form cannot move through it. So don't worry, it's not all size . . .

From this Density Of Dimension Equation a very interesting confusion is generated. Number of Lines Of Force and Cruxes does not per se decide how many Dimension's, only the Surface and/or Circumference of the Object decides how many Dimension's, one can easily draw as many internal Lines Of Force and Cruxes as one wants.

This functions on Real and Virtual Level's so it can also be called Real or Virtual Density Of Lines Of Force.

There can be Infinite Lines Of Force and Cruxes.

So, what are Infinite Level's of such complexities. In this sense the entire region of Space Time is a continuous tapestry of Matter and Energy's interwoven in varying intensities.

Since Dimension's have certain and uncertain Real or Virtual Density's based on quantity and quality of Lines Of Force and Cruxes which interact through Timelines then various Dimension's are NOT seperated from each other; just a note for those who already read my FREE Draft's, this is the opposite of what I originally thought, but over Time and plenty of Virtual Experiences to date I saw the flaw in my two equations and some other points which are now resolved. We are NOT isolated from each other, but now not because of some Abstract Moralistic Philosophy only, in fact how else are we supposed to get out there to even another side of the Galaxy: Everything below Light Speed takes a half eon. The varying virtual Density Of Form's harmlessly and sometimes quite not harmlessly pass through and around each other, like an EM Field.

Thus, various Dimensional Being's follow Law's Of Planes, Law's Of Dimension's, and Law's Of Quantum Physic's and Law's Of Physic's. One of the points of this essay is to expound exactly such since many now are convinced there is Not Only Classical Newton Physic's; it's more of a question what Dimension you exist in.

One Dimension or Plane merely seems insubstantial to the other, and thereby all Reality's, Dimensions, Planes exist simultaneously through and around each other.

Evolution

Cruxes and Lines Of Force can also form a Sphere. Cruxes being Point's Of Intersection form varying Symmetry and Assymmetry of Lines Of Force.

These Cruxes are bonds, Point's Of Intersection's, of Lines Of Force.

To achieve a Sphere, imperfect Spheres, streams of bonded Matter, go looking for other Matters, bound by Lines Of Force.

We are lattice works of partially formed Spheres, imperfect Spheres, in some cases no where near like Morgan Freeman's Science Series about the 2D Reality.

Through Evolution varying Lines Of Energy, streams of Matter, spiral around in chaotic madness.

Eventually Lines Of Power intersect with Lines Of Energy forming Lines Of Force converging and diverging at Cruxes in the Form. Eventually there becomes a closed circuit, where a series of Cruxes link, though this could be a Key Clue to Higher and Lower Dimension's (you can also stick minus signs in front) for 4th Dimension in Crux Diagram above looks definitely Leaky Wiki BG. Matter then continues to move, expand, spiral, circle, achieving more Cruxes. With increased Cruxes, more closed circuits, links, are achieved thereby producing more Circles connect to the Original Circle, more Spheres connect to the Original Sphere and so forth until a Perfect Sphere is formed. Of course, if there are too many collisions and conflicts then the opposite could happen.

Thus, an increasing number of Dimension's and Planes are formed and Matter works towards the completion of the Perfect Sphere.

As to whether it is necessary to become less dense, purposeful play on words, is still highly debatable across binding and dissolution of i.e. Karma, Sin and Relationship's.

This process is governed by above mentioned Law's.

I'd like to call it Law Of Unification (all bodies) but it's more like Law Of Bondage or Law Of Dis-Unification or Law Of Absolution these days . . . Anyway, as they say I can only propose something so you'll have to Vote on that one. [(;-)].

The Fact You Lack It, You Are Attracted.

The Perfect Sphere perfects all of Nature's Laws.

Truth: The Perfect Sphere embraces all Form's within it, thus holds all Truth within it, Truth being a Form. A Form is true unto itself and Relative to certain other Form's.

The Sphere holds all Form's within it, allowing for radius, diameter and circumference, so it holds all Truth, not just let's Circumvent Entire System again . . .

Cause and Effect: Holding all Form's in it, it is the realm of all Matter. In the realm of Matter, Form's affect each other. So, Perfect Spheres hold all Cause and Effect.

Do What You Want As Long As You Harm No One: Nothing penetrates an infinitely dense Sphere, thus a Black Hole at End Of Timelines, it penetrates Nothing. The Perfect Sphere is its own, no longer taking from other imperfect Spheres.

Repulsion/Attraction: Holding all Form's in it, it is the perfect Form. The Perfect Sphere holds no partialness, no Power and/or Energy bias, is perfectly symmetrical. So, any influence on it is Reflection Deflection Backfire.

All Things To Balance: The Perfect Sphere is a perfectly symmetrical Being, is the most balanced Form in Existence.

All Things Are Physical: The Perfect Sphere is the culmination of all Matter's, holds all Form's, Matter's within it, so is the perfect Form Of Matter.

The Spiral: The Spiral is a Form. The Perfect Sphere, the Universe, holds the Spiral in balance, by directly opposing, + and—, two Spirals against themselves. See Spiral Galaxy'ss. This usage of the Spiral helps the Sphere Form and maintains its structure. See Self-Consciousness, Law Of Unification (all bodies) front cover and back cover diagrams. Two equal and directionally opposing Forces create an almost unbreakable Balance and bond. It can be seen why the Atom which consists of Proton's, +, Electron's,—, and Neutron's, 0, is so difficult to destroy; it utilizes this principle. Unfortunately, now with proliferation of high-grade nuclear materials it seems not so difficult anymore for practically any Nation to do so . . .

The Circle: The Second Form. The Circle also helps maintain the Sphere Form. The Perfect Sphere essentially holds an infinite number of Circles in it, so you can imagine the Power And Energy it has.

The Sphere. Yes, the Perfect Sphere holding all Spheres within it, and being a Form itself, it is a super continuum of perpendicular equilibrium. Thus, the Sphere is the simultaneous One Form, maintainer, of the Law's and the forming, Creator, of the Law's. It is self-adjudicating.

Perfect Sphere: Nothing can destroy a Perfect Sphere. Thus, GOD, Infinity and Immortality is Not Impossible.

This great work of Evolution of Being's is to become just as perfect as Our Creator = Nature = One Big Reality, however for a Final Paradox we cannot = Our Creator.

We will copy for a Near-Ad-Infinitum, or even an Ad Infinitum, as we are products of Nature, for the Universe reproduces itself . . . Genetic's now proves this to us too.

Nature is a super continuum of perpendicular equilibrium.

Laser Military

The Laser Military is the replacement of standard Mechanized Military. This will take place in the future and hopefully first by Modern Western Civilization's. The priority is Defense not Offense, the goal is Near-Infinite Defense and not Near-Infinite Offense due to the result in History Of Humanity by excessive expansionists. The focus is Global Unification rather than racial conflicts. It is also necessary in the future to allow for the advent of Alien's; presently we have No Defense against Alien's despite the warnings of countless Science Fiction, Techno-Thriller, Fantasy and Horror Film's and Series.

As to my own Real Authority on the matter here now, next to the fact I am Sir Lord Prince Silber, Psionic Warlock at a min of 5-Star General as a Character Class: See my Microsoft Certified Systems Engineer (NT MCSE—All 6 Modules) acquired at TLS, Rodorgroep in Rijswijk, Netherlands and my further Full-Tim Paid Employment at Defensie Telematica Organisatie, thus the Ministerie van Defensie, thus in English the Information Technology Home Defense and support of Military in i.e. Bosnia 2001 of Netherlands. See curriculum vitae. Since I have about 156 versions of my curriculum vitae in circulation on Internet you might not have the latest version at my two websites. Next to this I am now also a Webdeveloper with many websites under the belt and my own custom Content Management System, the latest and greatest being The Open Markets with Dirk K. Blom, my Boss. Next to this and other things on my curriculum vitae I have found out about 2 months ago that I am a direct descendant of an Admiral Lans, my middle name 'Lance' was translated by my Mother in Toronto, Canada at birth, who was also raised to Nobility, the Naval Battles he must have seen . . . This makes me, Mr. Kyle Lance Proudfoot, effectively through bloodline, not only Rank, Status and Job's, at a min of a Lord Captain which my brain is just still not registering; if Oom Han had told me a long time before instead of what is looking like one of those Bwaindead Double Blind Britney Spear's Family Award's and/or a 04 Decade Family Initiation Rite since I turned 04 Decades on 24-02-2014 then I would have never honorably quit DTO because my Boss got hit by a truck on his scooter. You have each asked yourself why I hate scooters so much . . . I would have also never been such a Crazy Canuck and smoked so much pot and done so many stunts and gotten myself incarcerated on more than one occasion in a Celebrity with a Habit And Tradition Kick Off Clinic, a Mental Institute at Albardastraat called before Parnassia, now called Parnassia BavoGroep. It wasn't a total failure here, Life Is A Sine Wave, for it lead to Klein Westland and me meeting Dirk K. Blom who recruited me into Hollandse Markten which now has about 14,5 million hits. My own hits are since 1997 when I launched Silverlingo.com now 100+ million in total across entire Planet Earth. Unfortunately, no one can add all the hits up anymore through all the mediums. So, you also asked yourself why I don't tolerate anymore your dissin' and insultin' of me: I am not your Walking Talking Punch Bag and I am most certainly not one of your Blow Bimbo Media Puppet's.

That is actually a perfect expression for here since it is how a lot of Country's and City's look to date throughout History Of Humanity, you have No Defenses, the Culture who did not defend its Art's And Sciences was also ripped off and/or wiped out.

Oom Han, R.I.P., his real name I can now reveal if you don't know it already is Ir. W.I.J. Lans and his Father was my Opi Lans who Commissaris der Koninkrijk der Politie der Nederlanden in and after WWII. He was one of the 7 Commissarisen in Indonesia who fought against the Japanese and Geurillas in the Jungle and gained the title because he fought a whole group of the Enemy off after his Boss the Commissaris at that time got macheted down and died later in the Hospital. He was later captured along with the rest of my Family including my Mother who were thrown into the Japanese Concentration Camp. At my Mother's request I am not allowed to divulge any furthe details though we do have some written notes from their memories. He functioned as a Military and not a Police. Later back in Netherlands after years of unemployment he became an Inlichtingsdienst Officier in ILD if I remember the correct name for it cause I am now a BVD Officier myself and am appalled at the Holes And Leak's in IT and lack of Privacy's. These two things alone Double Compound Cascade Effect the problems these days and Chink Bitch Hole Hacker's, which is a Class and could also mean any Race since they are Numero Uno Hacker's on Planet Earth, and Hostile Enemy Terrorist Hacker's utilize these problems in our systems to no end. So now you know why I severely dislike Cop's, not because I'm also a part of Canadian Maffia since Toronto, Canada in only the Marijuana Department, after myself having also received €432,00 in Bogus Fines since Sep 2013 from Street Fine Agent's and other types of Neighbourhood Police who only locked me up to the date of 08-01-2014 for highly exxaggerated by your NL Media Poeperazzi Minor Misdeamenor's as what are now looking like setups in a background and/or in your faces Power Struggle; now I also have a new neighbour who is also in Police and at first we have not gotten along since we're both Aries, I may be Pisces in Birth Sign but my Moon is in Aries, but it's gotten better and it looks like we'll become more friendly with each other, though I don't think I need to describe and/or explain the differences to anyone between Military, FBI, CIA, KGB, ILD, BVD and Police . . .

I much prefer Energy and do not want your Throne, you could've just asked me too, which is perfect for the next paragraph since in the Near-Future there will be a breakthrough in the Energy Source for the following . . .

Laser Military consists of: Laser Pistol's, Laser Rifles, Laser Sniper's, Null EM Shield's, Null EM Spheres, Laser Infantry (Light, Medium, Heavy), Laser Tank's (Light, Medium, Heavy), Laser Artillery (Light, Medium, Heavy), Laser Helicopter's (Light,

Medium, Heavy), Laser Planes (Light, Medium, Heavy), Laser Fighter's (Light, Medium, Heavy), Laser Bomber's (Light, Medium, Heavy), Laser Space Planes (Light, Medium, Heavy), Laser Space Fighter's (Light, Medium, Heavy), Laser Space Ship's (Light, Medium, Heavy), Laser Mother Ship's (Light, Medium, Heavy), Laser Father Ship's (Light, Medium, Heavy) and the necessary Laser Cargo Ship's (Light, Medium, Heavy).

For more detailed descriptions see The Free Show—Version 4.4—FREE Draft and Planes Of Existence—1st Edition—Published by AuthorHouse UK which is the second part of my Science Fiction/Fantasy trilogy. The first part is The Black Dungeon Doorway—3rd Edition—Published by AuthorHouse UK. The third part is The Door Of Light or Door Of Light which is still what I define as a FREE Draft, Final Editt (purposeful misspelling) or First Run Of Final Edit. Apotheum Colluseum—The Ultimate InterActive™ Game—2nd Edition is the Battle System of The Free Show. The Free Show since 16-12-2013 has now become Spy Kill's 02 which presently has 4 FREE Draft's.

One of the most important things which Oom Han taught me, which now looks like an Officer's Training and not what we were told a European Education, is to first define all your objects, words, sentences and terminologies otherwise no one has any clue what you're talking about; you, I, me, us and them are definitely Caught In A Prism Of Your Own Design.

Unfortunately, next to all primarily disliking and/or hating each other's guts due to History Of Humanity to date we have relatively very little Armor's in beginning of 21st Century considering the Modern Warfare which will develop to Hyper Modern Warfare in Laser Military's and they have none. This is why Laser Military condescends from Orbit all the time by cloaked Elves on the rest of you and you each keep thinking at even 18 again that you meet the Laser Military Prerequisites. Well, no, the Laser Jughead is not the Jughead and i.e. if you do not have at a min of 150 IQ Level then you do not enter Laser Military Academy. This object is, of course, inspired by Jedi Academy which is a lot of fun using 2 Light Saber's and not only one. However, Planet Earth is not a 3D Game and often I think that Teenager's can never discern between Fantasy and Reality so I could also put at a min of an Age Requirement on it plus many other prerequisites in a Priority Per Sector Specialty System. Those are obviously not my Laser Military Department's. I, Sir Lord Prince Silber, Psionic Warlock, am the Chief Commander of the Laser Military. As a good joke, 'Do not try and contact me the Laser Military either!'

Thus, also appalling are the relatively No Security's and No Armor's on Planet Earth. The key to our survival into the future is to not let our City's get run over and not just by outdate Immigration Policy's. I'm not Racist, it's more of a Specieist, but if we do not update and upgrade our systems we'll have the equivalent of a NT Defense Hack again.

Next to Null EM Shield's and Spheres which are profusely described in my trilogy various other Types Of Armor like Steel Titanium Alloy's, Teflon, Chill And Thermal

Armor's and many other Element's CANNOT be ignored. Once we have other Colony Planet's and/or Mining Colony's we will have in the Near-Future an almost inexhaustible supply with Laser Cargo Ship's, not without Laser Escort again thank you very much . . .

As to Lies And Rumor's about Telecommunication's Industry that they are rapidly running out of Precious Metal's for Smartphones and Mobiles and as to whether China, a Stereotype Scapegoat, actually own 99.8% of all such Resources is exactly where I place such: So many bullshit and do not stick numbers on A4's.

A Laser Weapon is a powerful Weapon which can blast through anything. Laser Military has the following Laser Weapon's in 3 Global Category's which are greatly expanded upon in my other works mentioned above: Light Laser's, Medium Laser's and Heavy Laser's. The rough estimate on a Heavy Laser is a beam 2 meters wide with at a max Range of 2 km on earth and much further in air and ether space which can blast through anything near-instantaneously. The only defenses against Laser Weapon's are powerful Null EM Shield's and Spheres. For some of you who have read the FREE Draft's of this Evolutionary Essays you can see I have now updated and upgraded my Object Name Convention which I have applied in the second part of my trilogy.

Other Laser Weapon Class And Category's in Laser Military are: Laser Pistol's, Laser Rifles, Laser Sniper's, Laser Gun's, Laser Artillery, Laser Cannon's and Laser Turret's. I, myself, like the most Laser Artillery for they are powerful, energetic, fast, small and efficient yet pack a punch, especially like Napoleon you line up 1000 in a row and vaporize the Enemy.

'Sometimes the Best Defense is the Best Offense.'

'The Best Defense is always the Best Offense.'

I am still of the opinion though that it is erroneous and futile to go on Far Out In Left Field Nam Campaign's which only drain our own severely and generate severe hostilities.

Types Of Ammo are: Bullet's, Beam's and Blast's with ALL Element's.

Laser Military is the best complement to Free Democracy, 'Poetry Ad Infinitum, Defence Ad Absurdum' and 'Without a great Defence there is no Offence'. This are, of course, inspired by Island Mentality's like England who only survived thanks to Boudicca and Queen Elizabeth due to such and rightfully can poke fun at us back for Island Continent Mentality's, though as a Celtic Christian myself I could remind everyone next to the Power and Energy Lines how much larger our Continent's are. I'm still trying to figure out if I am only Scottish, English, Canadian and Netherlander or all of the above of North Europe for I do not lack German, French, Irish, Scandinavian, Swedish and Norway plus others in my bloodlines, associations and connections . . . And don't forget American too since Proudfoot Clan emigrated there to Louisiana in 1777 CE . . .

Following are the reasons why Laser Military has to replace Mechanized Military and will so inevitably though we are presently not a threat at all. As I stated in other chapters here and in my other works the present Laser Beam Prototype Fixed Device takes still a very long time to go through thick cement which is completely useless on the Battle Field. Also how, once again, does anything Recharge while somehow everyone in Compulsive Obsessive Disorder's can Regen across their Stamina Bar in 2 seconds flat. By my own estimations with all the Red Tape and Corporate Constipation for it would at this time in Evolution of Humanity retrograde and antiquate every last Weapon, Bullet and Armor on Planet Earth, not excluding ICBM's and Nuclear Weapon's for we simply Deflect Reflect And Backfire them off of Null EM Shield's And Spheres into their own populace, they are now the actual Villain and not us who do not lack our Right's Of Self-Defense next to Right's Of Defense and Right's Of Home Defense. Your whole Planet Earth Arsenal actually does not even get through our Titanium Steel Armor's without heavy sustained fire and since we are a hell of a lot Faster Stronger Smarter Better than anything and anyone, qua Heroes And Villain's, you will never in any Battle Scenario be able to achieve such sustained fire. To back this up with a recent reference to a Top Hollywood Blockbuster Film called Ender's Game by Orson Scott Card only in an extreme End Game Scenario with a whole Planet Invasion could you achieve such which is only about .000000 00000000000000000000000000000000000000000001% of all Battle and War Scenarios. It's, of course, not per se exactly that number but just to fill up the line an make my point.

Laser Military also researches, tests, shows and develops necessary improvements for Villages, City's, Country's, Continent's, Planet's, Solar System's and Galaxy's which are badly needed for Total Lack Of Defenses which some say most suffer from, thus the Deterrant Factor's to date, well if we have Near-Infinite Defense with Null EM Dome City's with exception of Null EM Train Supply Line To Polaris Scenarios then we can give them both middle fingers. It is primarily motivated by Battle Scenarios and Scenes portrayed in present day Science Fiction, War, Techno-Thriller, Horror and Fantasy Film's, Series and 3D Games of which I have seen and played very many; I think my present total is now 9000+ over about 03 Decades which is a rate of 3 x 3650 days = 10050, see I've never had any Math Bwain Cell's and fail that one myself even across the Real Host Mind and Body since it took me a half minute to calculate, which results in an average of not even 1 per day so that is not unrealistic; I could argue it is Per Per Only since does Com Department need Math, but I mean if you do not have at a min of Math Prerequisites somewhere there then everyone will still just go Dumb Stupid Blonde Platinum Alert!

Laser Military is a Need, Want and/or Recommendation for Individual's, Government's and Corporation's of the future, not the at Minimum Specification's Lag Effect's which most of us still badly suffer from. This plays a much larger role than you think with Bloated Impregnated Romulan Warbird Symptom's and Syndromes for if you're already paying 215% extra across Outdate Hardware and Software across multiple Laser Military Department's across Just Logon And Not Hack across DOSS

Attack's across RAM Bang Your Interfaces then you might as well go into Donation Fund's. Click. Boom.

In summation the following Principles of Laser Military give evidence and proof of the great necessity for updating and upgrading our systems:

1. Null EM Shield's and Spheres for Defense.
2. Laser Weapon's: Light Laser's, Medium Laser's and Heavy Laser's: Laser Pistol's, Laser Rifles, Laser Sniper's, Laser Gun's, Laser Artillery, Laser Cannon's and Laser Turret's: Bullet's, Beam's and Blast's with ALL Element's.
3. The inferiority of Police, Military and Mechanized Military with mere Weapon's, Bullet's and Armor's on Planet Earth, not excluding ICBM's and Nuclear Weapon's, do penetrate our Null EM Shield's and Spheres plus your whole Planet Earth Arsenal actually does not even get through our Titanium Steel Armor's anymore for heavy sustained fire cannot be achieved and maintained.
4. The lack of Individual, Government and Corporation Armor and Protection.
5. The brutalization of Individual's, Government's and Corporation's.
6. The barbaric harsh surroundings of a hostile environment.
7. The primitive sad and angry Colonial Education System's.
8. The lack of facilities, food, water, heat, cooling in their so-called Liberation, Liberalization, Liberosity and Leasure.
9. The lack of correct Port's and Protocol's to Protect Privacy and Security.
10. The instillment of incorrect discipline based in violence and not order; how many times will their Right Of Peaceful Protest be turned into a Violent Protest on purpose by the Power's That Be so they can Reverse Order Of Event's and Blame Each Other Again? This has also not failed to cause the Vicious Cycle Of Conflict.
11. The rampant Emotion dominating decision making and not Logic, Reason, Rationality, Argument's, Fact's, Statistic's and References.
12. The blood, gore and violence which prevails upon entire History Of Humanity.
13. The poor, unclear and ambiguous Statement's of Order's by any Department.
14. Obedience based on fear, anger and superiority instead of respect, courage and compliance.
15. The acquisition of useless Sand Territory in Far Out In Left Field Offenses. This one has actually not failed to cause Downfall Of Modern Western Civilization and especially Capitalism due to Excessive Cost Of Military and Extreme Enemy Animosity.
16. The lack of a Top Priority on Right's Of Self-Defense next to Right's Of Defense and Right's Of Home Defense developing to Near-Infinite Defense with Null EM Dome City's with exception of Null EM Train Supply Line To Polaris Scenarios.
17. The natural superiority of the Laser Weapon able to penetrate anything, even thick steel in minutes at present Point of Evolution in Timeline of Humanity, with

Null EM Shield's and Spheres cannot fail to grant Laser Military Superiority in both Defense and Offense so it will not fail in Global Union which does not fail in World Peace allowing for exception of an Alien Invasion who could themselves be vastly superior.

18. The need for Globalization, not repression, domination, elimination and annihilation.

19. The focus on Weapon's of Mass Destruction when such in not needed or recommended to conquer an Enemy since they also destroy the Environment.

20. The total disrespect for Life and waste of lives for useless territory and then withdraw and give it back to them again.

21. A Soldier who cannot be hit by Mechanized Military Weapon's. Thus, also Robocop's, Robot's and Cyborg Soldier's to conduct many tasks.

22. The lack of IQ Level's and Relative IQ Level's due to Outdate Colonial Education System's inflicting lag and damages on Individual's, Government's and Corporation's.

23. The low level of the population recruited into Military, Mechanized Military and Laser Military for numbers do count.

24. The poor slordid living conditions of Bunk Bed's = Bunker's and everyone Snore Into Insomniac Surreality's with no Serotonin so you really will enter Halluci-Nation's.

25. The reliance on bravado and morale resulting in massive casualties, "Charge!!! But not with the Credit Card, honey . . ."

26. The ridiculous quantity and quality of Money wasted on futile endeavours which Never Earn A Single Red Cent leading to National Debt's and collapse of Economy's.

27. The supposed call for Security, Serve and Protect with wasted Offenses, no Defenses and no Target The Cause For Once Complexes, "Conquer Mars, idiot."

28. A great lack controlled, clean conditions with Discipline and Order, with Peer Pressure and Leader Group Bashing, not to mention sharp-shooting . . .

29. The excessive quantity and quality of drugs, booze and boot-legs plus the obvious Sex Drug's And Rock And Roll in what are supposed to be ordered Regiment's.

30. The total lack of Normal Sex though it may be hard to define for some one can still argue the opposite across Extreme Sex since the 60's though some say it originated a lot longer ago in India . . .

31. The excessive usage of outdate and/or damaged vehicles, equipment and hardware, "Get some fuckin' Volunteer's man! Or what are you, a bunch of Mercenary's?!"

32. The lack of Command and Leadership on many Battle Field's due to useless outdate Com System's, over-complicated Rank and Department's, Cross-Species, Cross-Race, Cross-Culture and Cross-Language problems.

33. The denial of International New's Medias, Local Press, Free Speech and Information to the Public by the vast majority to date of Dawn Of Internet.

34. The targeting, victimizing, hijacking, kidnapping, killing, murdering, assassinating and slaughtering of Innocent Civilian's.

35. The prerogative of a multi-way Cross-Lateral Global Union and not Only Domination.

36. The lack of prevention and prediction of Alien Invasion's, potentially in futures.

37. The so-called Independence Day Debates, Battles and War's by Federation and Confederation covered by corruption. No one is fully independent or isolated by definition and fact, otherwise no more International Trade And Commerce and talking about Missing The Tourist Potential Completely again.

38. The excessive and fatal usage of Infantry Regiment's as Cannon Fodder when Air Superiority and/or Medium—Long Range Weapon's can take them out ahead of time.

39. The total lack of Short, Medium, Long, Space and Deep Space Range Scanner's with exception of only a handful of Spy Sat's, thanks to Maya Spy Sat again.

40. The mass delusion of Necessity Of Warfare and Necessity Of Evil For Warfare to date when Economic's through Trade And Commerce is always possible. Though Embargos are used to enforce various Majority Rules in United Nation's and now Group Of 7 it is not recommended for it leads to further Total Lack Of Bi-Lateral Cooperation's.

41. The lack of protection of Volunteer's, Laborer's, Journalist's and Investigative Officer's who trying to help in Natural Disaster Region's have to dodge bullets.

42. The ridiculous and inadequate treatments of Victim's, Prisoner's, Prisoner's Of War, Hostages, Kidnapped and Casualty's. The Violation's of Interrogation Processes And Procedures of Suspect's which amount to more like Torture and not Question anymore. Is there any respect anymore for Geneva Convention on Planet Earth? Are we also no better than the Enemy?

43. The strange outdate Habit's Of Regiment's, formations and excessive various Left, Middle and/or Right hand positions and salutes still pre-dominating everyone and everything which does not help our differences out, rather than War Games.

44. The excessive usage of swearwords, bashings and beatings in Military Training.

45. The wake-up procedure when you can set the alarm 5 minutes before they walk in . . .

46. The focus on Race Conflict's, rather than Specie Conflict's, when a Global Unification of Military Forces is needed and a global multi-racial, all for one, one for all Humanity is needed. It appears it is definitely not wanted by some.

47. The reliance on Verbal Order's and Command's at all levels when the only thing clear enough is written with wireless PADS through Near-Infinite Encryption of the Laser Military, otherwise one can just rip a 'not' into the statement, 'Ok, I'm

now going to give you your Order," says Silber, Psionic Warlock, "Do, not, attack North Korea! Hey, stop doing that, you naughty boy, you . . . Ok, again, Do (not) attack North Korea, hey, fuck you asshole, don't rip 'not's' into my statements!"

48. The lack of Modern, Advanced and/or Hyper Modern Entertainment for all Officer's and Soldier's such as Civilian Film's, 3D Games, Betting Houses, Sex Houses and so forth, 'Instead of bwainwashing them every time why don't you bwainstorm them . . . '

49. The excessive Prosecution Punishment And Prison for Act's of disobedience through excessive violence, beatings, disciplinary actions and very long Military Jail Sentences; you could also send them to a Front Line if they're that good at such as a Promotion: Facebook hires those who hack the fastest and best.

50. The nihilistic belief of Infantry in dying when it's Fight For Freedom and not just self-suicidal trench or mob offings and sacrifices; it actually states in this interpretation to defeat the Enemy not commit suicide unless it's strapped with C4, spring and jump-slide underneath tanks in the Korean War after WWII.

51. Our lack of belief, evidence and/or proof of the After Life through Atheism and now Anarchism on Internet, Nicht Ontheil Science and Schimdt, which gives greater strength to other Nation's who never dropped such; it also tends to lead to a manic depressive careless attitude with insufficient respect for Sentient Life.

52. The Military Training on Static Target's rather than Dynamic Target's in live real-time War Games of all kinds like 3D Games, Virtual Reality, Martial Art's, Laser Games and Paint Games; in Canada we actually started out in pre-school with Paramilitary Training and plenty of Fitness and Sport's all the way through Hyvescool.

53. The massive waste of Money, Resources, People, Vehicles and Equipment of Military and all other Sector's without sufficient economic payback i.e. Oil and other Resources And Territory. How can we protect them or us if the Oil Prices and National Debt only increase; we are better off developing Fission Fusion Ionic Null EM and/or Warp Engines a.s.a.p. away from dominance of Fossil Fuel Age; this is also good for Space Tourism and Space Travel since as I stated in a previous chapter here the actual logistical weight of the Fossil Fuel's and lack of maneuvarability is prohibitive; it is also advantageous for Home Defense and Consumer's as we step-over to hybrids to half the consumption rate across population explosions in no time by modification of cars; it is, once again, technically not their fault when People themselves have 2.6 cars per Family.

54. You are better off not on Only Welfare at Home and Home Defense, you could even Volunteer for Laser Military on entire Planet Earth which is its present form with no Leader to the whole thing and different Sector's talking to each other or not at all, you could also enlist up to a certain age and get a good to excellent Education out of it while helping your Country instead of Only Sob And Complain. The only thing saving America from a Ground Invasion is their 98%

Gun Possession, in Europe we have none again and only Hard German Steel Kitchen Knives; I, myself and me have 56 Kitchen Knives, so don't believe Lies And Rumor's you hear, we are not unarmed . . .

55. The failure of Psyche, Psychological and Psychiatric Test's, 'Sir! They're all fucking nuts. Some Order's are just insane!' They're also quite easy to fake and bullshit your way through. Actual Blood Test's and DNA Test's are needed with a better analysis of Array Of Symptom's Syndromes And Causes, 'It hurts right here Doctor . . . ' says Kid to Doctor, 'Where?' says Doctor to Kid, 'Here!' Kid point finger at sore spot, Doctor pushes on sore spot with right index finger and Kid flies straight into ceiling and keels over dead. Holy Gruesome, Spy Kill! This also leads to long term costs, no cures and exarcerbation.

56. The Majority of acquisition of Military Troop's is from lower levels of Society with none or low Education, which is still stuck in Colonial Ages for many, low or average IQ, and low or middle breeding. Because of their net worth each the higher Classes are even forbidden these days to enter Military's due to the Life Risk Issues i.e. Enemy Terrorist's and/or Soldier's even target Colleagues and Family Member's.

57. You see it in War Film's, all the time, yet you get do nothing about it, there are not enough Military Personnel being recruited in entire Modern Western Civilization due to the death of Nationalism and Capitalism with no Conscription and No Budget Crisis . . .

58. The trend of lost War's and recall. History Of Warfare Of Humanity proves an augmented cyclical repetition of the same errors and 'lessons not learned in this one are doomed to be repeated in the next one'.

59. The pointless gaining of Flag Position's when the territorial rights are national and/or international, except apparently for the Jews. This is hypocrisy. This has been the whole time a Not Unsympathetic Gesture for how do they not lack the rights to regions conquered by David and strengthened by Solomon? Yet, Germany got back all of its own Territory like many other Country's to date.

60. Ridiculous and stupid camouflage, which is still clear as day to see by other spectrums Gamma, X-Ray, Microwave, Infra-Red, Infra-Blue, Radar, Sonar, Radio and Bonk In The Night, which is all too expensive, which for some reason to date because of Animal Camouflage the White = Snow, Black = Night, Taki = Jungle, Brown = Desert, Blue = Water or Air, with a superior Military Force protected by Signal Jamming, Encryption, Shield's and Spheres one does not need camouflage, with Spy Sat's you're Not Invisible and/or Hidden: If he/she is color-blind, he/she can still see and hear you; you may not have Line Of Sight but Line Of Hearing is way more deadly, 'Oops, you brought a gun to a knife fight . . . '

61. Conflict's between Allied Soldier's is not to be tolerated. Yet this is rampant in all Western Military School's, like the Tumbler Boy who just because he has ADAD and not enough AD&D—2nd Edition, gets forced into Private Military Camp's and

bullied quite badly by being thrown into clothe driers. The whole reason the East is now taking over Planet Earth in beginning of 21ˢᵗ Century is primarily due to all the in-fighting in the West. You may have a different Heraldry, Clan, Tribe and/or Faction but we are on the same side of the War against Terrorism and those who oppose all Democracy's.

62. Extreme Punishment's, like no food, help nothing. Isolation Chamber's are now also proven to be mostly ineffective for Torture and/or Interrogation. The natural rebellious Spirit of Human always comes back up, that's even why many joined the Army.

63. The promise of Artillery Technical Education which means you get to be sent to the Front Line and all you get is Artillery, no real useful Education, Social Insurance and/or Family Payment to find a job or go on welfare, instead of Homeless Veteran's again, when the War is over again, you're cycled out, injured and/or sent home in a coffin.

64. Still too many Stupid Order's across Stupid Devices pre-dominating as there is No Budget Crisis, like to charge past a Machine Gun when you first need immunity to the bullets, it is still phenomenal to me that all Infantry in WWII had no Body Armor's, except how did the Evil Undead Nazi's not fall down from multiple bullets again, because going all the way back to first iron working they put on whole Metal Shield's under clothing . . .

65. Forcing Army Combatant's to remain with threats of Breach Of Contract, when an unwilling, slave, Not Voluntary Army is none at all, no better than Communism. Motivation versus Alien Forces is the key, not blind obedience.

66. Where are the Medic's? How long does it take to get to a Medic? There is hardly any Battle Field Medicine, 'Medic! Medic! Krrthunk, large serum needle, ahhh so much better, get up again!' I developed a 3D Level for Enemy Territory: Wolfenstein in .pk3 format and that is still my Numero Uno Favorite Element of it next to, 'Ahhhtacck! Ahhtack!' and since I was a Top Candidate with De Mazzelaar's who later went on to become Numero Uno Champion in the World I also like, 'Aaanvullahh! Aanvallah!'

67. Battle and War Field's is what Laser Military is all about, not only in 3D, 4D but also in 5ᵗʰ Dimension and up and down and around we go in Planes Of Existence. However, do not forget all Open Warfares and Debates which we see go across our computer screens on a daily basis now . . .

68. Laser Military is meant to improve Military's and Mechanized Military's of all Country's of Planet Earth, not to destroy it, with Numero Uno Focus on integral Defense, not Offense, with the long-term goal to develop Near-Infinite Defense for EM Domed City's, therefore those who are in violation of such as also stated by United Nation's and now unfortunately Group Of 7 instead of Group Of 8 are perceived as Enemy's, 'Always make more Friend's than Enemy's.' How else will we survive into the future and develop Planet Colony's, or versus Alien's?

69. Is Human and Sentient Life valued, or is it guns and numbers only?

70. If you are laughing hard and seriously taking this shit then you know you are suited for Laser Military. I was thinking of making a webpage or website with a Sam-Like Finger Poster pointing at you with the two words: Enlist Now!! But really, Laser Military at circa anno 2014 CE is starting to be actually developed: REMEMBER: In Guinnes Book of World Record's there were 2 Bogey's shot succesfully out of the air by 2 Laser Turret Platform's in Jan 2010 CE. It's still one of my favorite paperbacks on my bookshelves.

71. Fight's, Battles and War's between Allied Forces after WWII are never tolerated anymore, unfortunately though there are still many incidences of Friendly Fire.

72. Making an Enemy is not better than an Ally. There is, however, the necessity of Self-Defense, Home Defense, Defense and Near-Infinite Defense in the long-term. With the heavy reliance on offensives to date in History Of Warfare Of Humanity, like in Chess if you have no Defense then you cannot go on an Offense, we might as well play Exchange City's and Reverse Immigration every decade or so.

73. Valid Target's are Tactic And Strategic for a better Humanity, ' . . . like can we pls colonize Luna and Mars already again, tks . . . ' and not only one Individual, Race, Species, Government and/or Corporation.

74. The present Military Range Of Vehicle Displacement at only 50-100 km/h or mp/h is not even a moving target for Spy Sat's and make more noise than an old worn down washing machine with no rubber left banging on its last hinges and screeching hi-pitched scratching sounds and then knocking a cut hole in its plastic encasing sending 600 liters of water down on top of the neighbor's heads at the previous address . . .

75. The sudden introduction to hot, cold, misty, mountain and many other Alien Environment's when training is needed badly. Not later. Try no oxygen. One of the best James Bond Jokes to date is still How Long Can You Hold Your Breath? and is highly unlikely to be ever Bumped Off. I also like this one for all those who think they have what it takes, Battle Field Experience and are not just a FPS Couch Sofa Chip Eater: Get a hard metal object like a frying pan or baseball bat and hit yourself once, not too hard but just hard enough or you could hurt yourself, on the head. I go how did they do it to date?

76. Laser Military retrograde and antiquate every last Weapon, Bullet and Armor on Planet Earth, not excluding ICBM's and Nuclear Weapon's for we simply Deflect Reflect And Backfire them off of Null EM Shield's And Spheres into their own populace, they are now the actual Villain and not us who do not lack our Right's Of Self-Defense next to Right's Of Defense and Right's Of Home Defense. However, the Laser can theoretically penetrate ANY Target once there are breakthroughs in Near-Future in primarily Energy Sources, recharge and sustained fire potentials . . . there are also the actual Hyper Modern Material's

needed or the whole Laser Rifle just melts. Laser Military is the future of Space Conquest in the name of ALL Democracy's and those who are Not Enemy's of Democracy's for ALL People and Sentient Life on Planet Earth. The exception of Alien's still holds since they could definitely want all our Resources and Territory.

77. The reliance on very limited Senses when Scan Devices are needed for all troops.

78. If you can't pay for War, why are you conducting it?

79. The lack of Women in Military's, Celt's and Amazon's and others never lacked any, when so many Cultures in History Of Humanity also utilized Women for many other Support Task's such as also breeding the next generation of Soldier's next to knitting, sowing, building and chatting; back then they easily had up to 12 births, how else do you think they were able to do continual successive 10-Year War's . . .

80. The lack of focused physical training rather then sheer agression! Fitness and Sport's now teach and train for even World Olympic's with Brown Muscles. Martial Art's in the East have always trained focused Discipline with primary focus on Self-Control and not just show up as a Crazy Wild Jughead and start blasting away.

81. Information is the key. Knowledge = Power. Absolute Knowledge = Absolute Power. However, what everyone keeps forgetting with all the Racist and Reverse Racist Shit flying back and forth these days is that this does not apply only to Power which is now self-evident in History Of Humanity. You could also choose for Absolute Enlightenment with Absolute Energy and others, otherwise it does not fail in Gang Bang Effect in return.

82. The poor rush Short Range Tactic's when Long Range Scan's can first be used.

83. The underestimation of Enemy Fire resulting in too many casualties.

84. The susceptibility of Allied Weapon's with relatively no Armor's to date.

85. The overdramatisation and heroism leading to more unnecessary sacrifices.

86. The lack of compliance, cohesion, co-ordination and completion of 1 Clean Machine.

87. The reliance on hopeless Rescue Mission's and extraction when the original Mission should not have failed resulting in even more casualties. In some instances you can save a Fallen Fellow Fighter but what we've all seen to date the vast majority are not.

88. Laser's, not only Laser Targeting which Military and Mechanized Military do not lack, can improve accuracy and rate of fire with help of Scanner's, Spy Sat's, TRM's and lighting up, stripping and wrecking Shield's and Spheres.

89. A not-sure-win situation is no-situation at all!

90. Primitive Order's to 'move out' or Move! Move! Move! even shouted out loud to date when more Enemy's and Trap's are waiting is an outdate. Where are Laser

Scan's?? A standard argument by many these days is that we need a Laser Grid and not just an Infra-Red Grid to stop ICBM's.

91. Stupid self-sacrifice when team work is needed.

92. The lack of Kill, Murder, Assassinate and/or Destroy in Warfare as a Crime, Act Of Terrorism, Mass Murder, War Crime and/or Genocide. Where do you place the number at: 1000? 20000? 100000? To date there have been some documents and definitions made such as various Insurances and Right's Of Retaliation which are a great read but come across more like Term's & Condition's of Social Medias.

93. Useless Jungle Warfare if such regions are Not Tactic and/or Strategic.

94. The exchange of Types Of Prisoner's, even Lower Rank, for lives.

95. The abuse of Types Of Prisoner's when such helps none.

96. The lack of rations and provisions on Soldier's, try light Astro-Food! To date everyone is asking who is really winning the War when another $80k Soldier With No Helmet Armor gets sniped again and the $100 Hairy Terrorist jumps again behind his sandhill and into his sandtunnel.

97. The lack of in-field medical facilities.

98. If Warfare is too expensive then don't conduct it! The complete ignoring of Home Defense by everyone to date with the exception of only Obama who also got shot down on that one. Only due to AI in Pakistan did they get Osama Bin Laden.

99. Ridiculously stupid Night Missions, I mean talking about B—Stupid Violent Black War Film Humor, do you have any Infra-Red or Infra-Blue Laser Guided Night Goggles??

100. The strange demand for sleep depravation in long shifts which is supposed to make you more hardy, well No Serotonin Insomnubalic Surreality's is your brain chemistry and has very little to do with how hard you train for such. The only way for this are Pill's.

101. Don't laugh too hard, someone might hear it . . . Don't light up a cigarette, someone might see it . . . This is still WWII B—Stupid Violent Black Humor.

102. The inadequate, inhumane prison quarters, not to mention, God no, the treatments.

103. If this sounds like a total denouncement of useless pathetic ineffective powerless Military, Mechanized Military vs Laser Military, then you're right, but it will upgrade and update over time

104. The lack of scopes. All 3D Games laugh this one out.

105. All Minority Group's do not lack Right Of Participation in Laser Military.

106. What is more powerful and energetic then Null EM Plasma Laser Light, like the Sun, Nicht Ontheil RA!!

Each of the numbers can be perceived as a year starting from beginning of 20th Century when Mechanized Military thought it would be the best.

Conclusion: This list is by no means complete, even General's can add to it with their own experiences, I wouldn't go fucking with the numbers too much though . . . Soldier's in the Battle Field can add to it too. Anyone with at a min of Prerequisites Per Specialty of Laser Military can also add to it. Look at it as One Big Open Source Project.

The purpose of Laser Military is to provide an ordered clean highly intelligent efficient fighting machine for Near-Infinite Defense of Humanity and Sentient Life and the acquisition of more Extra-Planet Earth Resources and Territory's and to eventually also acquire Extra-Planar Resources and Territory's.

Only an Alien Species could oppose us or conquer us in the Future!

In our Fight For Freedom everywhere will we ever reach lasting Peace?

Origin's

In the beginning there was Nothingness, only blackness . . .

From the Big Bang, all things would proceed out equally from the Origin into Nothing which is outside the expanding Universe for otherwise by friction it would burn up. This is like one of those really bad Coffee Cup Effect's which they keep causing across Blinded By The Light unending: No, the Universe has its Origins in Black Hole Theory and Shadow Energy's which are more commonly known as Dark Matter's and Dark Energy's; it's practically already proven by the fact that the Solar Sun's are formed by hurled out radioisotopes. Egypt and other Religion's also have this described in symbologies: Osiris is the Father of Ra.

Zero Point Field Theory does not per se disagree with this for it says there is 'nothing which is not filled up 'in' or 'within' the Universe. No one really knows what is outside the Universe . . . it could also be only Shadow Energy which does not suffer from the resistance of Light Energy, thus the limitations of Mass to Speed Of Light, Black Holes are also such, just holes or rifts in Space Time and you could theoretically go there near-instantaneously. These things I have already discussed in previous chapters and other parts of my works.

This Nothing would have to be filled based on the Law of All Things To Balance with something, it could also be Light Energy like in a Platonic World where only Perfect Form's reside in a Light Field.

With an invasion of Matter into Nothing, the Matter would instinctively 'Conserve' itself in an act of 'Self-Preservation' and fill the Nothing: This also follows Law Of Conservation Of Energy. In the Universal Sense there is still a fixed quantity of Energy, so one bit of Matter/Energy moves into the created Nothing and another bit moves out. However, here we encounter a conflict again within Expansion Theory which states that the Matter/Energy is therefore *increasing* in the Universe . . .

It is possible with Black Holes for Matter/Energy to move in and out of Dimension's and Planes Of Existence and other Reality's. If the Universe is continuously and at a constantly accelerating rate expanding into continuumly, there has to be some balance in the quantity and quality of Matter/Energy or you get another Black Coffee Effect across Infinity, instead of being swallowed whole in the end it whites out in a flare. Black Holes could be sucking Energy/Matter into another Universe, thus by definition one which does not follow the same Law's Of Physic's or other things yet still has Matter/Energy, otherwise it does not exist again and is only Nothing.

Is it possible this Universe shrunk back into Black Holes? Is a Big Implosion possible? What is now clear, thanks to Spy Sat Maya again, is that Nothing has been confused with Nothingness for a very long time since first days of Hinduism and all the blow hurl suck mistranslations and misinterpretations. That is the Ocean of the Universe.

Problem: With an external Nothing existing, for something to fill it to the outer shell of the explosion, the outward spherical radiating Matter/Energy would have to be penetrated by Nothing. In other words, the surface of the Sphere would have to loop in on itself, the Matter sucking in the Nothing. This would then create an imbalance on the surface, an event horizon, as Matter pushes without resistance into the great Nothingness. Nothing would have to persist as Matter pushes effortlessly through it.

Well, look at a Null EM Field or a magnet which each Object and Subject has again.

GOD is in and about all things. Could this Nothingness be the Body of GOD, a kind of medium for Universal Consciousness? Thus, the Universe is the Body of GOD. It is after all, everywhere. Light and the Spirit Light is second only to Nothingness, filling it throughout the Universe. Third to that The Holy Ghost is Shadow Death Energy which sounds really Evil but in the trilogy of Christianity it is the third one. It is fairly blatant it was suppressed like so many other things since Humanity is not ready for it which I still think in my own Celtic Christianity is still a great retort. No, Noobies, there really are Big Bad Evil Insect's on those Planet's in another Galaxy even or the other side of our own and they are so huge they could crush a whole City here in one stomp . . .

From where does all the Matter/Energy come from except compounded multiple Black Holes? Say, sufficient Matter/Energy was sucked in to fill the Nothing. It pulls it from somewhere, one Universe leading to another as in String Theory. However, there is supposed to be a fixed quantity and quality of Energy/Matter in the Universe by Law Of Conservation Of Energy. That is thus erroneous and/or misapplied since otherwise we have to throw out Expansion Theory completely and state that it just exists forever going back and forth between different states as one Cannibal Galaxy swallows up another one and so forth, just like Andromeda and Milky Way Galaxy are moving towards each other and not at all expanding away from each other. Yet, how could their two Black Holes not pull towards each other and devour both of them?

Taking in the whole picture, another compensation has to happen somewhere. Let's see, something being attacked by Nothing is not going to be too partial. Nothing sucks in Energy from all points around it at once to fill it up, 'Nature abhors a vacuum'. Matter sucks in Nothing from all points on the Sphere as it expands. However, Nothing so very badly confused with Nothingness does not exist so that is still bogus: NOTHING DOES NOT EXIST for the last time again! However, Nothingness being Ether does exist. See also Greek Philosophy with atoms, Æther and what was word for 'breath' again?

Are we made in the image of GOD, does the microcosm reflect the macrocosm, maybe the Universe is the Mind of GOD, just like Planet Earth is the Mind or Body of Gaia, Mother Earth, just like The Moon is Mind and/or Body of Luna, sry cannot resist the Black Humor on that one, and Solar Sun is Mind or Body of Jezus, the Son of GOD.

That, unfortunately, make the Black Hole, how could they have missed the huge object off of port or starboard each and every other day for so long, you know who . . .

In Roman and Greek Religion's and Philosophy's, in Egyptian, Celticism, Hinduism and many others the Planet's are represented by the God's and Goddesses. Here we have again that phrasing and order of words problem: Don't you mean God's and Goddesses represented by the Planet's? It's quite phenomenal how they also line up across the Element's: Aries red and firy God of War, Venus hot and steaming Goddess of Love.

What happens to the surface of the Universe then? Apparently, it is expanding all the time, increasing the quantity of Energy/Matter of the Universe, which defies the Law Of Conservation of Energy. One of the most viable answers also supported by Stephen Hawkings and Morgan Freeman is the Universe is getting extra Energy/Matter from somewhere, this is most likely through Black Holes from other membrane Universes in Multiverse Theory: A. They have Black Hole Network's B. They touch at some point with each other at their surface membranes with osmosis interrelations and interactions.

The surface is not static. This at least we know for sure, or do we . . . That's part of the point here, no one has been there, scanned that far, the last frontier, the event horizon of the edge of our Universe . . . In all the photos and video I honestly see nothing but blackness which is presently being called by Astro-Physic's an Absorption, Obscuration and/or Occlusion Effect due to Light Absorption and too many Object's in between. Have a look at how many Star's there are in one Space Photo.

Nothingness is not static. It is infinite potential: Potentialus Ad Majorus. It is what Kinetic Energy and waves also pass through. Drop a stone from the top of the Universe and it has either Infinite Potential Energy or only Near-Infinite Potential Energy.

Don't take a slice of my pie. It pulls back. The result is a continual back and forth tugging, worse then two old Grannies bickering.

So we could start taking more seriously a perpetually existing Universe which only changes states and forms which is also the only Universal Model which truly allows for Infinity and GOD, otherwise GOD also gets sucked up by that One Last Black Hole!

'Y.O.U. are in Violation!' I am just going to love saying this expression every day too.

'A = A Violation!' Heh heh, not that I'm overly religious, these are Philosophy Essay's so through a cross-comparison and reduction we found this basic self-contradiction.

A Black Hole also contracts and expands. Each time a greater surface area is being pulled on; the surface of a Sphere pulling on all directions, while expanding, is greater than its Origin. Each time Nothingness is pulling on the additional external Energy's/Matter's, as well as the internal.

And there it is, there is no external Energy/Matter, except through Black Holes, or otherwise, invisible Portal's to another Plane and/or Dimension . . . Yet, we have already seen in a previous chapter here Planes and/or Dimension's do not need to feed on each other and are not separate, they all overlap throughout each other.

There is a fixed quantity and quality of Energy/Matter in the Universe, or better said, in the Multi-Verses, the One Big Reality, GOD, for otherwise GOD is only Near-Infinite and Finite being forced to fluctuate all the time too. GOD can only be INFINITE, otherwise GOD is Not Immortal, Omniscient and/or Omnipotent. A = A Violation!

This is also allowable within the Eastern construct of an Infinite Reality filled with countless macrocosmic and microcosmic World's and Universes, an infinite limitless boundless Nothingness, or is that once again the Nothingness *outside* and *inside* our perceived Universe? With the Multiverse Scenario it would also have to be. With the definition of Omniscience the only candidate filling the profile is Infinite Nothingness with very many Universes in it.

How many Universes, therefore? Our Modern Science is preoccupied with numbers and this is a Valid Question.

Matter/Energy is Finite and Nothingness is Infinite. However, Light Energy and Shadow Energy are also Infinite, some say even other Element's are also forever and everywhere. Are we, therefore, doomed to Extinction or only Transformation of Matter's and Energy's with Spirit and Soul in the end? Is that, though, not just another transmutation, trancension and/or transmigration which never ends or exits Reality . . .

There is not a fixed quantity or quality of Matter/Energy in the Universe. There is a fixed quantity or quality of Matter/Energy in Multiverses! For One Big Reality to be Infinite it also has to have a Constant Infinite Matter/Energy/Spirit/Soul. Otherwise, with expansion, life and growth you will also have compression, death and decay. This is acceptable for a Mortal but absolutely not for an Immortal.

Throughout entire Reality, which just is, there is Nothingness. Light can reach even the most dense of objects like a Black Hole for they have defined it as a Near-Infinite Point Of Singularity, I only use my own Name Object Convention, and what is one the most famous statements by Christianity, 'GOD is in all things, each and everywhere and in you, even the most Evil of them all SATAN does not lack one glimmer of hope.'

Whereas, it may take billions of years, if not longer, Near-Infinity, our Galaxy is doomed to die, our Universe is even doomed to be destroyed, however one cannot say that of the Multiverse, One Big Reality, which exchanges Matter's/Energy's/Spirit's/Soul's through multiple Universes all the time. You merely choose by your own thoughts, words, actions and deeds how long you want to spend in a Hell and/or Heaven Plane.

Let's have a look at our own Universe. With continual compensation needed on the sphere of the expanding bubble, the Universe has to pull on it's own Energy/Matter.

The result is a continuous cycle, a continuum, of the Spheres, Energy/Matter pumping through the Origin, with an ever-increasing quantity and quality of Energy/

Matter being pumped to its surface. Thereby, an ever-increasing radius and circumference with a continual flow of Matter/Energy cycling around and through the Origin, the center point, the beginning of the Big Bang, or the climax point of the Big Bang. So, technically speaking I am not disagreeing or disqualifying Big Bang Theory, it has its context in our Universe. However, with such it is bound to contact another Universe at some point across Billard Ball Universe and Membrane Universe and also Multi-Layered Universes and plenty of other theories to try and explain our nihilistic existential entrapment.

There also must be a center of the Multiverses, the Throne Of GOD or if it collapses back down to a Black Hole which was another Violation then SATAN again; otherwise, it does not exist, again, for ALL Object's have a center point. Regardless as to whether our limited measuring instruments can find it, like in Multi-Dimension Theory, as I showed in my previous chapter here they are very holy and assymetrical, an existing Object has to have Dimension's which by Law's of Mathematic's and Physic's must also have a center point, whether real and/or virtual.

Another way to describe this, is when one point pulls, it causes along chain of pulls, which when taking into account the chain of Universes, as purported in String Theory, which is a fixed quantity and quality of Energy/Matter/Spirit/Soul for the Multi-Verse but not only the Universe, will pull unto its Origin, the Great Origin, in other words, looping. This looping, allows for Infinity of Energy's and Matter's. So it occurs, ad infinitum.

Thus, Nature is a super continuum of perpendicular equilibrium.

Why perpendicular? This is still the key question of the day (07-04-2014—Planet Earth—Milky Way Galaxy) in Quantum Physic's and Quantum Computer's which should hit the market in about a decade only . . . Matter pulls in Nothingness from the surface of the Sphere. From any point on the surface of the Sphere to the Origin, the line will be perpendicular to the tangent of any point on the surface. The flow of Nothing is perpendicular to the Energy/Matter. This allows for the necessary balance in all systems, 'ALL System's work towards a state of rest.'

Nature is a continuous cycle of Nothingness, Energy, Matter, Spirit and Soul. This happens with Everybody, Everything and Nothingness, all the time. This is what sustains the Universe, stability by the cancelling out of two opposite Forces, Nothingness and Energy Matter, a SCOPE.

Space Time is therefore a curve, as it exists within the Multiverse, the Super Sphere or Umbrella Timeline or Pi Race Course or many other Timeline Theory's. In any case, they are all curved, 'There is no straight line in reality, except a perfectly virtual straight line from A → B.'

There would be no stability, everything would fly up apart in the worst Chaos Theory Effect imaginable, we do not exist, we are only virtual and nothing would be bound.

'Everything needs Nothing in return . . . ' We have to define things by their opposites because we are bi-polar, so now I throw a Curve Ball Effect at my own shit again.

Each particle is doing the same process. Each is given the same command, the same original cause, to continually fill the Nothingness.

Unfortunately, this still does not give the cause or reason of the existence of the Multiverse which is now high abstract Philosophy. Who or what caused Infinite Reality?

It is too difficult to sort through a cause of a cause of a cause etc. and it is unimaginable to think Nature/Reality did not always exist, somehow for the worst meaninglessness all gets destroyed again, has no original cause, was never begun even, then just reloops . . .

However, this does show the end point is the same as the Origin of the Universe it loops through, the hermetical snake biting its own tail and the hippocratical double helix snakes and other not so pleasant snake symbologies . . . The Origin Point then becomes irrelevant as the continuum flows. All points on the Sphere can then be considered Origin points as they will all behave in the same manner. After all, each part contributes to the whole. With this each particle has a spherical field. And with such we can also open up a Mini-Black Hole anywhere and go anywhere near-instantaneously.

Even though this does not answer, 'What is the cause, purpose and reason of the Multiverse?', it however does affirm the Existence of our Universe. Now let's try not to go extinct in our Evolution for by the fact it exists it must have cause, purpose and reason.

We know we must go to Center of the Universe. I suspect this will go along the lines of personal Enlightenment, not by technology.

Well, such is the crux of the Matter . . .

Trends

Imagine a world where everyone allows others to change freely. This would have to be in a world where everyone is given food, shelter, clothing and the right to do what they consider interesting.

The whole world is confused on how to supply these needs and wants. Everyone thinks it is bound to religious, political and economical factors which for some reason prohibit the vast majority in partaking of both; there are apparently still well over 3 billion people on Planet Earth at or below the Poverty Level, many are homeless and starving as we speak. That is HALF of the population of Planet Earth! And some of us dare call ourselves Near-Enlightened Advanced Being's even because we have Internet. Well, even if that is their argument then give them Internet . . .

There is this desire, however, for unconscious flow. It is easier to go to University with your parents supporting you, even if you do not know what you really want to study. In fact, parents will often push you into this unconscious flow and how can you resist? All you have to do is study. Study, studdy, duddy, dildo, ' . . . and oh wait, I know now, now I want to be like a Ballet Dancer . . . now I want to be an Actor at YouTube . . . now I want to be a Jewish Soldier . . . ' Supposedly this study will give Independance. With tuition fees, these days, it does the opposite and you get caught up with the flow of Society, never truly realizing yourself, only working for materialistic means and ends which Society tries to enslave you in. Then you have to do the bidding of The Thanes and their Master's whipping you on in a fevered fervor making only more debts, ' . . . it's like some strange nihilistic phenomena where they go around in circles everyday making even more minus signs and debts, well which way are they going around the circle, clockwise or counter-clockwise . . . Holy Gruesome, Spy Kill's!'

What if your desires somehow contradict the methods of the powers which be. Like, Exxon, do not hate them, they're still pumping 2 million barrels of oil per day out of Alberta which is a consumption rate by the People themselves, they're no more an oppurtunist than anyone else, if you don't step over to Types Of Hybrid's pronto than you'll be toast in some decades only, if you don't stop being Stupid Dumb Tourist Consumer's then you'll be dead in decades only . . .

So do you go with the flow, or fight it?
What do you do when you are Superman?
Or have super goals.
Like no pollution. Food for all.
Understanding of all Space Time.

Exploration of all Space Time, without difficulty.

Infinite exploration of the Universe.

This means easy potential, no, zero difficulty in exploring.

In a continual state of exploration of all Space Time.

Now such is super goals. I believe each of us could do such.

So, do I give a damn about the world, or just my own goals?

What most people don't realize, is the two are inextricably bound together.

For sure, we won't be able to do anything productive without sustenance. So, it is up to everyone to figure out, to the best of their Capability's, how to sustain themselves and preferrably everyone through self-sustainment and self-preservation. Otherwise, you have increasing homeless, starvation and dying.

There is this thing about unconscious flow, the tendency to do everything the previous generation dictates, to follow through with what they have laid and improve on it, but not change it, this is the classic fear of change, of the Unknown, again, but now in beginning of 21st Century when we have recently found out with various Deep Space Scan's that we really are still completely alone in the Galaxy, no sign of Sentient Life near, we have to learn that this is our Only Barren Rock in Existence, if Planet Earth is about in the middle of Milky Way Galaxy and by Obscuration Effect the Alien's are on the other side of the Galaxy or even only 100 Light Year's away that is also how long it will take before we ever have contact at Speed Of Light.

Fortunately, in our cases, thus a small percentage of each Country to date which is steadily growing on Internet thanks to Free Information and Social Medias, this could not work: How could anyone follow in the path of present day self-destruction action of the Gas Oil Corporation's who want to keep us permanently in Fossil Fuel Age when there are a thousand forms of Alternate Energy's? I still like Solar Energy which is already Near-Infinite . . . Am I supposed to, say if I were the Son Of The Owner, or if even I am anyone, find better ways to use Gas and/or Oil and/or Coal when there are thousands of Types Of Alternative Energy's? Don't make us Laugh Hald!

Am I supposed to find better ways to use Nuclear Power?

Am I supposed to find better ways to use Hydro Electricity?

Am I supposed to find better ways to tread further on Planet Earth, to conquer it? Force our dominion even farther over the Earth. And better, yes, better! We must improve! Yes! Let's go to the Moon, no wait, such isn't far enough away from the Issues. Mars! Neptune! GlibetGlob Sector 3! Let's go there first with No Maneuverable Rocket's instead of developing Null EM Space Jet's, even Ion Engines and better Sources Of Energy from all things with better Energy Storage Cell's.

Am I supposed to Own Everything and practice No Guardianship with total lack of any self-sustainability?

Am I supposed to work towards Space Travel and Space Tourism and be the worst Blood Cannibal of 'em all living them all dead behind?

Absurd! Ridiculous! Self-annihilatory . . .

Meanwhile, how many Humans are dying, unrightfully, with no justice, here on Planet Earth as we speak? What's the Rate of Death : Rate of Birth and how much carnage in between. And you dare each ask yourselves why the superior Alien's have not showed up?! They don't want your Death Complexion's to spread to Reseller Galaxy.

Some say if Stupid Humanity does not stop on its present rate of course to nearing a self-extinction there will be a Mass Extinction Event like St. Andreas Fault Line. Why has it not happened yet? Yet, in the very paragraphs of Christianity itself it states, 'And then cometh a great earthquake which causes the Apocalypse' It then describes how this will effect each and every last Nation and Country on Planet Earth.

If I need to quote International New's Medias and statistics about the present rate of all the self-obliteration's everywhere then Look It Up On Internet again!

What is it about inverse reactionary relationships? The inverse Law of the Universe. The Law Of Opposites. The Law Of Newton. The Law's Of Einstein. The Law Of Schrödinger even does not fail to tell you each these things already. Nicht Ontheil Zen Buddhism!

Everything has an opposite. Everything reacts in an opposite and equal direction. Good, maybe it will work if you apply it for once to your Religion's, Politic's and Economy's.

See, it is possible to look at things in many ways. However, a rose by any other name remains the same. And a black rose by any other name remains the same pain . . .

Yes, folks, there is reality, if it was only Dream Reality then you would not die or go back and forth between the two. And let's hope it don't come too hard down on all the unconscious flowers with all their twisted broken stems, 'I am your Prince of Law, Light And Order and you will work the fields and we will not fail in selling them on the Market's, get up and work instead of staying on that Suck Welfare, or you'll also be stuck with 2-Bit Dry Cracker's, how do you think I got here to The Open Markets website on Internet? I did nothing but Free and Cheap Labor Job's . . .'

Of course, then maybe they we will all burn as the dirt collects in the fuel, a spark goes off and the whole thing goes up. The reason I state this is not because I am a Nihilistic Anarchist, but because each of us, you and them must Never Underestimate the Forces of Chaos and Anarchy, randomnicity has now played its role so many times and killed, murdered, assassinated, destroyed and wiped out entire Civilization's by just only Not Unnatural Disaster's that it will happen again and again . . .

If these Individual's, Government's and/or Corporations think they are in Power by i.e. central takeovers of banks, then they have no concept of the potential of the People and other Power's and Energy's who do not fail to see our Death Analysis Trend's such as the Grain Lord's and Utility Lord's. Now with Telecommunication's, Internet and Social Medias working together on a daily basis with International New's

Medias there are also a hell of a lot more who are working towards a better Future for everyone. Some say it's now in 2014 CE after the last worst suck year ever a ¾ to ¼ weighted system and not Spliffed Down The Middle anymore . . .

It is pure evil we have sent rockets to the Moon, conducted warfare for gas oil and not spent the money on renewable, clean and possible Infinite Sources Of Energy, Near-Infinite Sources Of Energy and Alternative Sources Of Energy; these are not Free Sources Of Energy, you still need all of the digging, boring, drilling, mining, hammering, screwing in of locations, installments, equipment, hardware, circuitry, local and remote computers, people, service, sales, distribution and then the same again for the home of the consumer. If they do not change over now and it comes down to the wire, there are going to be some serious repurcussions. However we know now thanks to a lot of AI and R&D that they are Stuck In Prism's Of Their Own Design and signed entire Modern Western Civilization down and up river with Signed Contract's for the next 100 Year's. The way we look at it now is that they have doomed themselves for we never signed . . .

Why is the size of Greece worth of Rainforest's which possess all the cures for each and every Sickness And Disease known to Human being cut down per year? You could also do Selective Cutting and Culling with Chocker Block's like English Farm's. Nicht Ontheil All Greanpeace Activist's all around the world, however if you do not get is signed at their Legislative Level all of your Ground Level Activist's have been for nothing . . .

We would all panic if our Analog or Internet Radio Station's and Television Station's gave the facts. But would we rather soak up the drivel of their Individual, Corporate and Government brainwashing influence on the media, 'Why Bwainwash them if you can Bwainstorm them?' Don't believe me? Who is paying them with commercials? These weirdos in corporate seats have this obsession with Power and determination of Reality. They believe they must do this for their Safety And Security.

You mean again, only *their* Freedom, Safety And Security.

Now, just say this isn't true. Then, oh my God and my Goddess, we are all inventing the whole thing and I am some weird, freaked out, backward, pseudo quasi neo Hindu Communist (Hippy) who believes in Only Ascetism, again, all the worldly pleasures are Illusion and/or Delusion, only the Soul and Spirit is important, all material Wealth is evil and I want to live on a not unhorny half-naked Commune, again, smoking Organic Pot Grown By The Fully Dynamically Dark Full Moon, again, and all their blow hurl suck attempts at Label Image and Not Impersonal Attack's, again, everyfuckin'day.com, again, I am not your Walking Talking Punching Bag, again, cause they don't have a single retort, counter-argument, foundation and/or fundament left to stand on with the Last 2 Burning Pillar's Of Society and Wibbly Wobbly World.

So you see No-Noobie and/or Noobie Chat Session's of The Free Show are useful for somethin' . . . I mean, really, no, man, your Opinion, your Not Unofficial Opinion, your Unofficial Opinion, your Shit Ignorance, your Bullshit does not, once again, decide or determine my Self-Image and does not Insult me, I really don't give a flyin' crap what any of you think of me, my works, my work, what I do or not, how I wipe my ass, all the way to the other side of your blow suck hurl Lies And Rumor's Hollywood, Amazia, Bollywood and/or oh oh who is now the 4th Player . . .

Enough already. What we argue is the acquisition of all the pleasures for the people, not too mention advancement of the conditions of Society, Races, Species and Humanity.

'Anyone who purports the above onto those who talk for improvements through change, as improvements come only through change, will end up in Hell. Or stay in Hell' Who said it this many times to date as their now having Panic Attack's watching their Empirialistic Empires crumble down to the ground again . . . Well, what the hell do you expect, it's no longer the Stone Age, it's the Ether Age.

If you are putting down a Minority Group, again, then you are an asshole, if not a Racist.

Yes, there is Hell, they're called countless Hell Planes now, also thanks to me and us who did plenty of R&D and now just like Microsoft they're having Jealousy Fit's cause we now also have a Bad Evil Monopoly, for Minority Groups too who conduct Reverse Racism, that was Not Unpredictable, for the Environment which has always wiped out Villages, for anyone, anybody and anything for such are Law's Of Universe which share Common Element's. Yes, there is still Wild Life, but if imbalanced why not a Cash Crop?

Planet Earth could be a very nice place for people. Presently, only for those who can afford it, who are fortunate enough and who desire it.

Maybe all the major City's will just collapse . . . Basically, no one wants to see it all collapse. People have worked inhumanly hard to achieve what we have in Modern Western Civilization's and there's too but those Stock Crashed are 2-Bit Obvious.

Celestial Forces

Plane(t)(s)

The following description functions on a virtual level of Reality, not a physical one. Plane(t)(s) start out as bonded particles which then cruve around to their beginning, through Lines Of Energy, at such a point the line becomes a Circle. The Circle then gains more particles, until more Circles at slightly different degrees cross over until enough of them form a Sphere.

Each Dimension has Lines of Energy which have circled around enough times to form a Circle: A Plane. Circular Dimension = Plane.

The location of a Plane, what determines where and what the nature of the Plane is, is in the exact centre of a number of intersecting Circular Dimension's.

The 3rd Dimension exists between 3 Planes as one draws a line through each Plane to intersect in the middle.

Thereby, a Crux is the intersection point of the Dimension.

The Planes act as the inhibiting Energy binding you to this Dimension. In Science this is known as strong or weak Energy, the one bringing together, the other breaking apart.

In the Universe, the Sphere, there is an infinite number of Planes. Thereby, once again, not only do Dimension's co-exist in the same Space Time, but all Planes co-exist, coalescing throughout each other. See also Chapter 24, Origins, here in this Evolutionary Essays.

Point = Crux is in a Line = Line Of Energy is in a Dimension = Structure Of Lines Of Energy is in a Plane = Plane Of Existence.

The Microcosm reflects the Macrocosm. However, does the Macrocosm still actually reflect or be influenced by the Microcosm as Theosophy purports through Eastern Religion's and Philosophy's for it is still hard to bend your mind around that such a tiny nanoscopic particle can influence such larger Celestial Body's. As Native Indian Religion's and Philosophy's state a stone cast into a pond casts ripples but how does a cute little fish here actually send bubble up to Pluto . . . Is the Force of the Fish not too small and must first get a lot bigger?

Now Here.

The Lower Planes have greater density of Matter Energy, so less Light. The Higher Planes have great Light and less density of Matter Energy, in a gradual kaleidoscope of worlds all around the brilliant central Sun's and Black Holes.

Sun(s)

A Sun is the result of the friction of intersecting Lines Of Force. See previous chapter here for a detailed definition for Lines Of Force.

The Crux is what forms the Sphere. This point has all degrees, is the point of Balance, and the resulting Friction, Fusion, of intersecting Lines Of Force creates an explosion of Light Energy into all directions equally.

'Oh my God, that must be quite a Blast Radius . . . ' 'No, got another one on radius to date, it has a Blast Diameter, now you know the one Not Untrue Meaning of Karma, for there is no Implosion and/or Explosion which goes in only one Direction . . . if you cause that kind of disturbance in that region than all the other particles will go and stomp it . . . '

Sun's are Spheres Of Light.

Simultaneous, there is a brilliant central Sun created by the overlapping of all the Planes. Some call this the Throne Of GOD.

The brilliant central Sun is the Origin of all Life.

The brilliant central Sun is the Being One With Nature.

As all things, Dimensions and Planes, exist simultaneously, Now Here.

As all Light Energy exists simultaneously, Now Here.

In the brilliant central Sun, the Origin, there is only One Law, Axiom, Crux, and this is Truth and Absolute Reality.

Existing in Truth and Absolute Reality is the Being of Light.

Fountain, which sheds a little Light on all Matters . . .

Nothingness

Ether has been perceived as Nothingness by the Ancient and Classical Philosopher's, now also most Modern Philosopher's and even many Literature, Culture, History, Religion and Science Expert's do not disagree with this anymore. Science has proven Ether is not Nothing, as stated in a previous chapter here the millennia long confusion definitely lies in Nothingness ≠ Nothing for Nothing does not exist. See also Zero Point Field Theory. It is just more subtle, like a thin gas, than the ancients could perceive for in their great Out Of Body Travel, Vision's, Higher Sight and Divine Prophecy and Providence they saw emptiness between all the thousands of Galaxy's which we now have for the first time, thanks again to Maya Spy Sat. It permeates the whole Universe, but it is not Nothing. Considering the new wireless microwave speeds and other wavelengths which go through the Ether what else would it be?

'A lot has been said about Nothing.'

Throughout Everything, the Universe, all Matter Energy, there is Nothingness.

However, Nothingness = Ether has to be in something itself too which is Light Energy which in turn still has to be in something else which Light Spirit and Soul Power and Energy which has to also be in something or rather Nothing outside of the Universe or else burn the whole thing up again by friction causing another god awful Coffee Cup Effect, ad majorus potentialus infinitum, causes all Motion. Matter Energy fills a Vacuum, Nothing, Light suffuses through it all.

'Nature abhors a vacuum.'

Who we are is based on an exact configuration of Matter Energy. Within all Matter Energy, there is a lot of Space, a lot of Nothingness, approximately 98%. This exact configuration is determined by an interaction of Matter Energy with Nothingness; Matter Energy is in continual motion to fill Nothingness to Nothing, to become more balanced and symmetrical. See previous chapter here on Dimension's.

Nothingness has, as of yet an Unknown Speed Limit for wireless wavelengths, Nothing has No Limit's. Since the motion of Matter Energy is affected by Nothingness to Nothing, a great infinite potential, the motion of Matter Energy has no limits. Expansion Theory has now also been strengthened and actually does work hand-in-hand with the others. See previous chapter here for the description.

Everything is Infinite Power And Energy of Nature, One Reality, GOD. Everything continuously expands into Nothing.

To understand this picture Nothing as what is beyond the Great Sphere of the Universe. Also, picture Nothing as Space with no Matter Energy in it.

Note here: Nothing is a part of Reality. Reality is Everything and Nothing. To describe Everything we agree that Multiverse is the best term instead of other spelling variations. There is, after all, more than one Universe, more than one Dimension and definitely a lot more Planes Of Existence than we thought. In Judaism the number 10 where 1 = Everything and 0 = Nothing also represents this. Also remember how the number 0 was imported from Babylon.

Within Everything, the Great Sphere, are the Form's. All of these Form's are limited. The Form of the Sphere, itself, has one limit: It's diameter. Beyond such is Nothing. Within such is Nothingness. One must first achieve the Form of the Sphere, perfection, before achieving No Limit, Nothing.

This is where the mistake has been made by Eastern Philosopher's.

The way to Final Liberation is not to meditate on Nothing. Or do Nothing. Or eliminate the Body, it is to perfect your Form, become One With Nature. This can, indeed, take many lifetimes, as Energy is conserved. For in the greatest Paradox of Buddhism how does Buddha not cause anymore Karma, thus even ripples would bind him again, if he still exists in Nirvana? The only way to not cause any new Karma, good, neutral and/or bad, is to not exist.

You have to perfect your Body, each of your Body's, the physical, the mental, the emotional, the astral, the spiritual, the light, the godliness (atman). Apparently, as according to Theosophy, seven-fold. This is to allow for the unobstructed flow of Nothingness through Matter Energy. When this occurs, when you ascend, when you have a perfect Sphere of the Body, your Form expands its one limit, its radius, and reaches to embrace Everything ever-expanding into Nothing.

I had this Debate with my own Father quite often when I was a strict Vegetarian in Toronto, Canada from 14-22 years old and I also disagreed with his Jainism and Ascetism and Not Violence and Escapism and Puritanism and Old School Pict and Victorian Francisism because he thought then, and probably still now, that it really is about dissolution and/or absolution of bound karmic particles in your whole Life, that you have to free yourself by deleting your relationships even . . . Well, no, you have to resolve your Issues, solve your weaknesses and harmonize your resolutions. Only then when you become a more balanced Sphere instead of Law Of Opposites and Duality where your Pendulum Swings To The Opposite Degree.

Say this happens . . .

Nothing goes beyond Everything.

Literally, analogically, figuratively, parabelically, symbolically, metaphorically. Most importantly, actually, really.

There is actual real Nothingness = Ether. Such permeates all of Reality for all the Galaxy's go through all the Universes which go through all of the Multiverse.

After all, Freedom, Peace, Near-Enlightenment, Enlightenment, Omniscience, Omnipotence and Bliss eventually become redundant, yes, boring. So, off to the ecstatic Being of embracing Everything, which only Nothing can do, with Nothingness in between all things, no conflcit, zzz, ahhh, finally, fire Sleeping Gas Canister's at them and once again the lion lies down next to the sheep, ah, so very Not Violent. Screeeeeech, record scratches to a halt, 'No, there is no such thing as Not Violence in the entire Universe for it was formed by every impact and collision imaginable, and even Andromeda and Milky Way Galaxy are on a collision course with each other, both being Cannibal Galaxy's.'

In fact, they are all Cannibal Galaxy's, it's just a matter of which one is bigger or which Black Hole is stronger and/or bigger. Just like Sun's they have Life Span's.

This is always the Now Here as we spiral through Time Space, which is a well known long-standing play on words with Nowhere and Here Now from millennia ago, a superb way of describing the Unlimited State(s).

Thus, where is Time in it all again? See my other chapter here about Motion for a detailed theoretical proposition.

Nowhere means no one particular point. Now Here means one particular point.

If you cannot be defined to one particular point, it means you embrace Everything, exist Everywhere, as you have no particular formation which would identify you within a Dimension.

The moment you would take on a particular Form, you can be identified, labeled, stereotyped just to poke some more fun at all of them in Hollywood . . .

The Form itself is the Point. To be pointless (hah hah) is to be Formless. Thus, Nowhere is perfect for Nothing, as Nothing is Formless.

Now Here, however, also equally describes the Now Here which is the Being of all Points, simultaneously.

So it looks like it means the circumference and center point at the same time. This cannot be, this is another Paradox, two different points cannot be the same point.

As shown, Nothing embraces Everything and Nothingness is in between it all.

Here Now is Being in the moment, the present momentum, Past, Present, Future.

Being in the moment and Present Momentum is noticing the stimuli coming to the senses of each of your Body's.

Celestial Body's also go through Time Space.

Now is a Time reference, without a Time measurement. Now states only the Present Momentum = Present Moment. It does not state anywhere 9876543210.0123456789.

Here is a reference to Space. Space states the present where XYZ RST Co-ordinates.

The previous description of Now Here takes care of What, Why, Who, When, Where, How. And, how the hell did we get here?

Thus, we have a book. See last chapter of The Black Dungeon Doorway—3rd Edition—Published by AuthorHouse UK.

Now Here works.

Human

'Oh, Immortal One, what is Death to you?'
'It is Nothing.'

Human consists of ALL Element's, however primarily Top 10 Element's: Earth, Water, Air, Fire, Ether, Shadow, Light, Form, Spirit, GOD.

Human Being is the only Being on our Planet to possess all Top 10 Element's in a Good, Neutral and/or Evil Self-Conscious Form, allowing for Spirit's who may still be wandering and wondering around to see how they died or have to find rest.

Human has plenty of Fire. Fire is also Energy of Self-Consciousness next to Light Energy and Shadow Energy channeled down from GOD and up from SATAN. You could say Fire casts both Light and Shadow, it is definitely used for Good, Neutral and/or Evil, and so is the nature of Human Being.

We are the Circle, we are working towards a Sphere, with potential Immortality.

Through ALL Element's, we are linked to all other Being's in Mother Earth Sphere. We are the most powerful and energetic with Capability's to use Top 10 Element's, with Self-Consciousness, or to misuse such with Unconsciousness. With this comes a reverance and responsibility to all brethren Life. Without, we die . . .

If Human continues to be so self-egoistical, arrogant and destructive with his fellow brothers and sisters, other Life Form's then Immortal Ones will come down and annihilate every single one of us and/or them.

With Globalization uniting all Country's, presently consuming all other Country's, it seems we will avoid mass destruction as one cause effects the whole. However, at our present rate of explosion of over-population and consumption the chances are slim.

Only Time will tell:

'Only mass destruction.
But that will not occur,
While there are still
Good people afoot,
Making things for each other,
Lending a hand,
Giving a smile,
Making love,
Working for the promised land.'

Meta-Science System

The system of Meta-Science (MetaScience or Metascience or Meta Science) is about Human Being, Humanity and Human Species; do not be a Racist and/or Reverse Racist.

This system consists of Observation and Experience, the Senses and Logic and Emotion.

Observation and Experience

Observation and Experience is the awareness of your Senses.

The way to discover Truth is through Observation and Experience.

Observation and Experience is the usage of your Senses to discover Truth.

Observation and Experience is not Science, Philosophy, Art or any other Human system, however it does incorporate all the fields into one Universal System of Knowledge and Application of Known and Unknown.

One cannot discover Truth if one do not both Observe and Experience. These systems are all the result of Observation and/or Experience. As a result they all hold parts of Truth but can't figure out the whole Truth for whatever problem they are working on. And all have become ignorant of, or have dismissed, the importance of the Senses having written them off as Only Subjective Reality and stepping completely over to data, numbers and Only Empirical Method's; don't forget that Truth Is Commonality and I can just compare thousands of notes with others. See previous chapter here.

Modern Civilization's, not just the West this time, are predominantly Observation with a severe lack of Experiential Knowledge having been told by Conqueror Rewrites History and this is more of the blame of Empirical Classical Science then Christianity even:

'And all the teachers in the world

Can't give the secrets of Experience.' Poetry Lore—1st Edition—Mass Energy, The Death-Life Conspiracy—Published by AuthorHouse UK

We are saturated with systems, techniques and ways of Observation, yet Experience and the Senses are seen as purely relative and subjective. This is the product of a primary male left-mind dominated Culture.

For Human to progress, Experience must also be learned from, otherwise a Being cannot fully understand its Human Condition; this can theoretically apply to Animal's and Plant's too if we even know their Behavior and Language correctly but is beyond the scope of this essay.

In effect, without Experience, Truth cannot be discovered. And, of course, vice versa.

Experience is no less fallible than Observation. Both are subject to Human Condition.

For example, being Objective during Experience is ultimately recognizing your own fractional Condition and can lead to contradiction with your conclusions of Truth.

The purpose of Observation and Experience is to discover Truth.

Since Human does nothing but Observe and Experience, I hereby state the entire purpose of Human is to discover Truth.

Every problem we have demands Truth to solve it.

Truth is the One which embraces All.

You can, of course, roam around in blissful ignoring of it all, however you are still experiencing diverse subjective interactions.

Thus, Truth is Human's Genetic Prime Motive (HGPM or GPM). You could also call it Growth Per Moment . . . You cannot hope to fulfill yourself in anyway if you do not discover the Truth of a problem, if you do not Observe and Experience . . .

Observation and Experience together is the only way to discover Truth. One can also argue collective Observation and Experience, not recognized by much of Politic's and Science, yet fully defined up to Freud and to Jung and now even Modern Psychology is throwing away all of our dark and/or fire psyches away making lame excuses as to the costs of such while there are no cures coming for the Human Condition, thus actually viscious looping everything and causing more costs, and doing en masse again from there unending empirical Only Neurological Biological Psychiatry. Yet each of our Soul's, Spirit's, Mind's, Body's, Logic and Emotion's are rooted in our poly-animalistic natures. Going also from top to bottom we are pivoted in Pluto as all spirals down to us from the Heaven's above and all the great Rationality, Reason's, Ideal's and Ideas. Modern and Quantum Science even can still not put their finger on it yet we all are made of such.

Since the two together embrace all things, when one is lacking, an integral point is missed, and resulting Tunnel Vision (TV), narrowness, and missing of the larger picture reigns. This is what I mean when I say on a daily basis now that you are possesses by Lies And Rumor's of Medias, Lies Of Satanism and your Shit Ignorance; if you do not know how, again, to cure yourself from Yellow Rot Butsky Fever, you'd be surprized how long it takes to grow all the way down your legs and up to your lungs before you finally die of suffocation then it is most certainly not Bliss. Nicht Ontheil John The Baptist!

Since all fields of Human endeavour have Observation and Experience in them, not denying the incorporation of any or all other systems, Human Being simply needs to embrace such a potential:

'Ode to be invulnerable, never to catch the thorns of fear.
Never to diminish in capability and potential,
Like an untorn child, open to the wonders of the Universe.'

Poetry Lore—1st Edition—Self-Consciousness, Law Of Unification (all bodies)—Published by AuthorHouse UK
A lacking, a con-fusion will remain.

Body—Senses

We can only understand things in terms of the Senses (all 6 of them): Touch, Taste, Smell, Sight, Hearing, Intuition. Lower and Higher Senses are theoretically not impossible to achieve, look at it more like Caracal Ear's or Bat Sonar or other Poly-Animalistic Capability's which are far better or fitter as in Darwin, and from top to bottom again some claim to have Lucid Dream's, Vision's and/or Divine Insight's.

Senses are integral to Observation and Experience. Yet, most of us, live in the realm of our thoughts paying very little attention to the details which pass by . . .

Experience is always greater, we observe just a fraction of the Universal Experience, the Now Here. The beauty of it is, Nature is all ways giving it to you. We simply need to open our Awareness, our Senses to it and the great Soul's, Spirit's, Mind's and Body's who came before us and are yet to come in the next generations.

The Energy Manifestation's of their Density, Composition, Formation, Resonation, Vibration, Oscillation and Radiation of particles and waves comes to us by reflection of Matter and Energy off of covalent bonds. Our Sensory Manifold then picks up this reflected Energy giving us impressions, interferences, disturbances, conflicts and very rarely do my neighbours ever harmonize with me in return.

Thus, our senses are reflections of Nature. Our Consciousness comes from the Senses as we formulate ideas based on Sensory Input through signals to the brain. Our Consciousness, thereby, is a reflection of Nature. Is this what could be meant by an 'image' of GOD, what if a mirror image is meant? We are, therefore, smaller angles to different degrees and the Angel's, God's and Goddesses are larger ones of GOD.

The brain can be seen as a lattice work of many Lines Of Energy overlapping. The flow of neural messages is fractionated Energy being transmitted from one 'node' to the next, one brain sector to the other. There are, of course, millions (if not billions) of nodes and are very small.

The refracted Energy then triggers the particular center, node, crux which is connected through Lines Of Energy.

Fiber Optic's is similar to this.

One center's purpose is for the maintaining of all the other centers with motion towards Balance, Conservation of Energy, Freedom, a Sphere; until then you are unwhole.

One center may, after input, resultingly shift itself in accordance with other centers within the brain and the body to achieve such a goal.

With Energy lacking, the center's cannot function, as there is nothing to relay on. Likewise, with increased Density, increased number of Lines Of Energy there is less Light Energy or Shadow Energy pouring through.

Matter produces Light Radiation and then such bounces off of other Matter's. Light does not penetrate Matter, it penetrates through Nothingness in Matter. See previous chapter here for description of difference between Nothing and Nothingness.

Thus, Lower Planar Being's are less evolved than Higher Planar Being's as they are more 'Dense'. Being more Dense is having less Nothingness within your Form for Light Energy or Shadow Energy to flow through. See also Solar Sun's and Black Holes in previous chapters here and in Science, Astro-Physic's and Quantum Field Theory.

Thus, neural impulses, transmissions of Energy, have less motion in those more dense. Think of fat and obesity and no fitness and no Brown Muscles; you think at first that Brown Muscles are more dense and make you more 'stupid' but the Brain is not the Body and strengthening of the spine cannot fail to help the Brain which then becomes more straight, better flow of juices, not bent or broken and therefore not blocked.

Ether in Beer, Wine and Alchohol help the flow to Brain as long as it is not Overdose; this varies per Psychology, Physiology and People for Frenchmen grow up with Wine and have Genetic Immunity, Scotsmen grow up with Scotch Whiskey, Russian's grow up with Vodka, Canadian's grow up with Beer and Wine, entire Europe grows up with practically every Type Of Alchohol and North America now is KABOOM!, thus as many of you already know Native Indian's never had any Alchohol and so it is Wild Fire.

Light as in Visual Spectrum Vision is one example. With other Senses the neural impulses are not that Type Of Light, they are billions of neuro-biological electrical impulses which use neural synapses, neurons with up and down states at a nano-level.

Evolution is about changing all of one's Matter into Energy. This way your head will start glowing. Is this is what is meant by the halo? Even better than a light bulb. So one needs to increase their points, their centers, their cruxes, their bundles of Energy. Lines Of Force are Lines of Power and Energy, streams of bonded Matter, and Motion is what is needed to keep your body which, fortunately, includes your brain (hopefully) working. See previous chapter here for definition of Lines Of Force.

The basic difference between Life and Death is Life = Energy and Death = Matter. You consume Matter and kill it to generate your own Energy which is your Life whether that be Alien, Human, Animal and/or Plant to any Degree Of Consumption.

So, fill all your GAP's with Energy, but try not to consume entire Planet Earth in one day.

Mind—Logic and Emotions

All Human innovation and invention comes from Logic and Emotion's. These are Human's two basic functions of the Mind. The right and left hemispheres and all the brain sectors work together to form Memory through learning with Observation and Experience, 'You can only learn through repetition.'

I will qualify them: Logic is used primarily in Observation and Emotion is used primarily in Experience which result in Unknown—Known = Knowledge. Look at your own Relational Argumentation's, Resource And Territory Conflict's, Debates, Battles and War's. Emotion's are exactly such E-Motion's which we each go through by such.

Knowledge also comes from Information which is stimuli, you need to do anything. See Heidegger Theory, Matrix and Number Theory's.

Emotion's are also Energy you generate in any action.

Logic is a comparison of points, dimensions, planes with virtual variables leading to Truth is Commonality; there are of course many Relative Truth's and Relative IQ Level's which is now a Mut Point. See previous chapter here for more details.

'Good Logic' is impeccable, infallible, unforgivable and absolutely undeniable.

'Good Logic' must not be tainted by Human's subjective rationale. How is this possible if Human is using it? Because, again, it has its own Law's. Mathematic's is actually one of the very few systems which practically everyone agrees is a Universal System = Objective System.

Logic was not invented by Human. It is a part of our Form. Our Form is a product of Nature = Universal System = Objective System.

Logic is, thereby, a product of Nature.

The Law's of Logic are those of Nature, as everything is in Nature, to give you another bit of Logic.

If you do not follow Nature's Law's you fail in even Logistic's, there's another.

So What?

You must use Logic to acquire Truth. Without Logic you do not know if something is True or not. How do you know what you are seeing is True or not? Well, you apply the Law's of Logic:

'Use logic to prove actual danger to you.'

Poetry Lore—1ˢᵗ Edition—Self-Consciousness, Law Of Unification (all bodies)— Published by AuthorHouse UK

The Law's of Nature which essentially compose Logic are:

1. Logic: Do not contradict.

This 1st Law Of Logic is a rephrasing of the One Great Law of Nature: Truth.
That's it!
To put Truth into practice in your life you need 'Good Emotion's'.
Do they have Law's?
Yup:

2. Emotion's: Do not contradict.

With this 1st Law Of Emotion we see in us the One Community, the Commonality: Truth.

Truth is the One Solution to every problem through Observation and Experience.

I could go through a whole List Of Law's here but that would take another 100 pages and is beyond the scope of this essay, 'Always leave something up to the Expert's.'

Sex And Interaction

Sex/Gender

Each person grew up as a Young Child, Child, Young Teenager or Teenager who's mind was on nothing else but sex. Their sex/gender therefore . . .

As a young one is continually conscious of their role or where they fit in this role in every way revolves around your sex/gender: What does a boy do? What does a girl do? What kind of boy am I? What kind of girl are you?

Sex is the base, the gonads, the foundation to everything else in the spectrum of a person's character.

When a person is known for a quantity and quality of character it is immediately associated with male or female, boy or girl, man or woman. When one is gentle it is either a gentle woman or gentleman. If one is strong you are either a strong woman or a strong man, in completely different ways. Though the word is the same, the association is different, you think of completely different things. In fact, every word is associated with male or female differently, just as defined in Languages. Try it!

It is natural. Our sex/gender is our basic difference. In fact, almost everyone, minus those who are aware of this, and then not even, go through life associating with everything in this manner! It is because it is the basic part of our Being's. How else does one make sense of the mirage of collages of Reality which is dualistic in Nature? See DNA, Chromosones and Spiral Galaxy's plus many other sources to back up Duality.

Associate. The primal way of understanding things. The premise.

This is why it is so brutally cruel and unfortunate if someone is not permitted sexual release, has their sex stunted. Sex is an integral part of our Being's.

To promote sex in this Culture I have a recommendation.

There is no good meeting place for sex i.e. a Sex Bar.

A place is needed where it is accepted by all the people going there, they are there for one reason, and only one reason, to have sex. In fact, it could be gays are ahead of everyone else in this aspect. When you go to a Gay Club you know you are going there for sex. So, if you're not gay, watch out!

This is needed for each Age Group starting at Teenager, 'Why wreck the teen? Don't wreck his/her Sweet 16, read the number, 13 is when Teenager starts, if that's not their number on the nose culminating at 18 and then post-blowing all the way thru the 20's.'

Single Bar's are a Myth Hit.

Bar's, Dance Club's, Techno Arenas, Metal Stadium's are not really for singles. They are places for couples to enjoy themselves. With any other place, including these two, such as a sports game, activities, school, work there is a lack of clarity of premise as to what his/her intentions really are, do you have anything in common, do you even know each other at all, do you have a common friend and do you know what the dangers are. If you walk by yourself into a bar you'll probably get kidnapped, hijacked, interstellar highway robbery and gang raped. Internet is no better these days, in fact a lot wors, since you can just lie, cheat, steal and hack all of yours and your victims computer system.

For example: Why does a man approach and talk to a woman in these places? Is he trying to come on to her? Right in the hallway? Is he just interested in 'friendly conversation'. Yah, right. Is he hoping to get information from her, like what time she gets off, where she lives and so on, so he can rape her? Yes, it is this bad.It's borderline sexual harassment just offering a 'Biertje' these days in NL with dubious oppurtunity.

The problem here, everywhere, is not figuring out what the purpose is, but is in the fact we are actually trying to figure out the purpose. There is no clarity.

The place where this can be solved if a Sex Bar is opened, not just an Internet Sex Bar, is one of refinement and taste serving all things to assist in the process, not get some horrible STD, SOA and/or die of AIDS while being ripped off for a half century per quarterly. The only reason anyone goes there is to get laid, and get laid real good. This would be publicly accepted, except for their blow hurl suck Outdate Politic's.

Wow, how kinky! A place where people are not looking for 'relationships', how gross, please not another Relational Argumentation leading to the so-called American Dream = Relational Enslavement where in the traditional Family Model the Father has to slave away 40-60 hours per week with 1 Wife, 1.663 Children, 1 Suburbian House, 2.13 Car's, 1.530 House Pet's, Front and Back Grass Lawn, White or Black Fence, 3.13667 Credit Card's, Sex Once Per Week and could you please go hang yourselves with your own rope, dope and tie it all up, I know it will soon be better, where no one is just there for a drink, where everyone is there for Only Sex. There is presently no adequate place for any sex/gender cause I wouldn't trust his/her photo equals a dog too as far as I could hurl throw suck blow their whole profile . . .

'Not for teenagers, 20, 30, 40 and onwards.'

'Sex Bars are needed.'

'Badly.'

'Please.'

Interaction

To further maintain Human sanity, interaction is necessary.

We don't interact healthily anymore, it's even all virtual with Multiple Personality Disorder's ala 313 profiles on Internet and counting, it's not in person like at least there's just the two of us with Spliff'd Down Da Middle and only 2 Schizophrenic profiles.

Oh, we talk to each other at work, but this is done with pre-programmed statements and answers, most of them geared to cease the conversation as soon as possible, without appearing rude, cause you're wasting my time and I have work to do (Shut up you _()_)(()*%$^%$%%^^%_)(_(_()_&%%$%^$^%%$^_)_)(())_(^%$everyfuckin'day. com).

Here are some more examples of daily banal boring conversation:
Example 01:

"Hi."
"Hello."
"How are you?"
"I'm fine, you?"
"Great."
"Great."
"Talk to ya later."
"Yup."

How else could it be? People at work are associates, not friends.

In some cases, the only thing in common is the job. And how much can you talk about your lame job these days?

Example 02:

"Hey, isn't that a fax machine?"
"Oh yes, indeed!"
"It's the latest model with super smooth paper injections and no smudges along the creases."
"Yah, let's talk about the phone system, now!"
I am risking my reputation, now, using this word, but I can't resist: 'NOT !!!!!'

And in some jobs, you can't get away from the other person. Imagine if you are a pilot . . . , 'Shut up. Shut up. Shut up! SHUT UP! Weeeeeehhh . . .' Kersploosh! Holy Gruesome, Spy Kill's. Did you know, once again, that one of the Top 10 Worst Statistic's to date is that Employees everywhere waste even up to 50% of working hours per day doing nothing but the most useless Bla Bla Bla . . . Now, do you also want to jump in bed with your colleague, the Nightmare Relationship Scenarios in Hollywood Film's alone . . .

So, to offset work, we go home, meet our spouse and our kids, and go to an iMovie.

Woh . . . First of all, there is most likely a huge argument about what Movie, then obviously the parents rule as 8-Year Old Joey, who wants to see 'Kill and Die', starts crying, then probably during the car trip the parents fearing a hugely embarassing situation decide maybe 'Six Degrees Of Seperation' is a little to advanced for Joey and decide on, yes, a Family Movie where all the animated Animal's blast by at Mach Speed. So kids actually rule. Of course, the intellectual cravings of the parents are now compromised.

Then at the iMovie, after balling his eyes out for an hour screaming 'I want my iPad!' repetitively for an hour straight, Joey wants to sit in Row 3. After all a kid is small and it's actually really cool to sit about 4 meters away. Unfortunately, there is no way Mom or Dad are going to suffer eye palpatations. So Joey starts crying loudly again in the bioscoop and irritates the living crap out of everyone there, now everyone's pleasure is compromised. Parents rule again.

And the situation is more tense than at work, and people go day in day out with this.

The mistake here lies not in these people's behaviours, they are trying to do what is natural for them, 'How much do we work to find rest . . .'

The problem is the choice of activity and the interaction and the elements and everything from cry scream holler Step A → Z.

The desire is to have a nice evening out with the Family and enjoy each other's company. You have to enjoy life. Sounds perfectly normal, right?

It is because the choice of activity of pleasure in no way accomplishes the premise.

You are not enjoying each other's company when you all sit in a row all facing in the same direction, each involved in the Movie. Or am I missing something? Sitting like a bunch of headless zombies, and anyway, why would you repeat such when you do such at work. Such will not accomplish your objective: Your inherent desire lies in interaction with each other; you think that is interaction but it is not. Another classic example is the Absent Father which if you look it up is another god awful statistic . . .

O.k., so you tack about the iMovie afterwards. In other words you rehash it and pick it apart until its torn to bits. And if it's the Family Movie there is enough substance to talk about for about 1 minute. And Joey is already onto other things.

What is needed is personal interaction. Face to face parlé, mutual commonality. Physical interaction, mental interaction, emotional interaction, spiritual interaction. And not fighting, cursing, spitting, lying, cheating, stealing, hacking, hissing, clawing, biting, ticking, hitting, punching, swearing, hammering, knifing, shooting, killing, murdering, assassinating and/or destroying. Look at statistics everywhere, this is presently dominating ALL of your Relational Argumentation's because of exactly one thing leading to all the rest of the miseries: 1. Pleasure Center in your Brain.

This is also your Reward And Punishment Center which is also a suck acronym.

From a Young Child to later as an Adult you have each been denied quite a lot of such and/or traumatized by multiple use and abuse instances, the common blame being Sex, Drug's And Alchohol, well, no, by the fact you got none you have to overcompensate, and quite badly.

This is already proven by Psychology, Psychiatry, Medical Science and many other official instances. Look it up, it's all their on Internet now, not just in Entertainment.

Interaction is done for the sake of interaction. And you will find all your desires satisfied. One thing is not per se 'bad' or 'taboo' or better than the other, it's all per genre and category now on Internet.

For example, a most satisfying, enriching, enlivening, interactive activity is jamming in Music with your Friend's with no intentions of becoming World Famous Rock Star's but then oopsy the Band Politic's take it all over again and wreck everything.

Why can't we just play music together like throughout entire History Of Humanity, now it's who's got the most hits at 'you know who', please for once without that gore Pop Media unending. You don't need to be an expert.Don't just watch TV or listen to Radio since such Not Internet Broadcaster's are State Run Propaganda Medias; technically speaking the Liberal Corporation Internet is the bi-polar opposite of State Run Analog TV and Radio.

Get a hand drum even. Drum's are the musical interactive tool for people who don't know what a piano is. Hand Drum's are the base, the tribal instrument. The heart beat. All people can relate to a drum. And if you get bored with your Bongo then you can always work your way up to a Drum Set and then Virtual Instrumentation.

It all rattles, hums, shakes and rolls.

Play with music or without.

Dance, sing. It is fairly common now to dance at a party.

But pls, and not sry, for once could you shove your whole f'in Politic's and other shit which wreck all the creativity where the Sun Never Shone.

Most people won't sing. So hum. Get together and make some music where each person does exactly what they want.

What do you think Native Indian's, African's, Celt's, Eastern Indian's and all other tribal communities, clans, communities and everyone to date did?

Modern Society needs a big injection of tribal interaction.

So why is it all fucked up now? Are we disconnected? Are we too over-complicated? Why can't we just get around a table or go online and play for fun in a Public or Private Group without everyone fucking each other up? Well, there's that god awful Money again which they all do it for, 'For what or how much money would you wreck someone's 3D Gamer in the last Battle Scene because you run out of Ammo and have to go back 12 Near-Impossible Degree Of Difficulty levels to figure out where you fucked up?' Get doomed you Chink Bitch Hole Hacker!

3D Games are meant to be excellent interactive tools for sexes, stimulations, simulations, scenarios, sensations and staples of fun. 3D Games emphasize the people, such as Classic Games like Charades, Pictionary, ' . . . and don't tell me my drawing is not 3D you fucking Bitch, that's all you do is wreck all the Art in it all the time!'

And there we go again in another miserable spiral of Stupid Relational Argumentation's.

Why do you think they are so popular?

Now to get really tribal.

How about stories?

Each person gives a 10 minute story which is intriguing to the others. The Natives did this profusely. Entire Literature is based in such 'primitive' roots of our Society's.

How about acting? Some people suggest a scenario and some others act it out, much to the hilarity of the Observer's. The Celts, Greeks and Roman's did this one frequently. Well, now the entire Planet Earth from Hollywood to Bollywood to Amazia is ruling.

A variation on Charades would be this: Though it involves three people, it is good. One person is 'Experience' who acts out something. One is 'Knowledge' who makes a statement. One is 'Observation' who must interpret the other two. The roles respectively state what the person is doing and understanding what the person means by their statement. The roles cycle around.

The game actually has a very great psychological potential in the communication field.

It is also personalized. The person who is Knowledge must keep in mind the person doing it. Otherwise, there is an incorrect interpretation.

No points are needed. The interactive activity is fulfilling enough.

The whole idea is to regain the reality of a Community interacting with each other. Instead of the isolated mentality and/or over-complicated nature of Modern Society where everyone has become a Mass Entertainment Consumer.

To expand upon this, as it needs to be done, simply take any activity and see how it can become more interactive, or how it can be applied to an interactive scenario, such as the examples above.

With this implementation will come a very powerful surgence of happiness, satisfaction, postivism and productivity.

In other words, Human sanity.

Not the perpetual continual Human Insanity's.

P.S., 'Fuck you? No, go fuck yourself, you're so much better at it!'

Money

Money

The biggest and if you come right down to it, the only worry of anybody in this Society is Money.

In this Society you need Money to survive and survival is a slightly important issue.

You need Money to do anything here. Money is a great tool, a totally objective tool. Netto 1 Dollar means the same to everyone, since you can buy the same things with it.

Everyone, unless they have an inhibition can become a Millionaire, just sell a lot of shit. You'd be surprized how shit sells like B—Horror Film's.

It will be great when the entire world uses one currency; Global Union, though not per se run by Global Canadian Citizen's, with other governing bodies like United Nation's is not a Neo-Commi and/or Neo-Nazi Planet, however my own concern which I brought up before here and in my other works is how do we make any Money at all if there is no Currency difference? As my Uncle David stated, who is an International Businessman who owns Result Clothing in UK, by the time it takes to fly back from China to UK he's already made a loss by the shifting Currency Rates since when you sell in larger quantities you often only make a couple percent profit margin . . . Most International Trade is based on Currency, this is how since Colonial Ages and millennia ago with the introduction of Coin by Caesar in the West and before him going back to Mongolia, Persia, China and Japan in the East, and possibly even back into unwritten history in the form of Gold, Silver, Copper and other ways to barter, we have made Money.

However, one of the causes next to excessive expansionary policies of WWI and WWII was the end of Colonial Age with begin of Industrial Age and the strong desire and will for Independence, Freedom and Liberalization from those who misused, abused and/or enslaved Colony Country's i.e. Indonesia and Japan.

Also we have now seen that Capitalism in the West has badly failed with out-of-control Black Hole Effect's, Negative Spiral Vortexes and a National Debt Rising Ceiling which is causing god awful Higher Inflation Level's. We forget so quickly the lessons of the past as that being the Primary Cause of WWII with the rise of Hitler in Europe. In North Europe we now also have very bad Inflation, once again as stated before here it costs per person per day €1 for breakfast, €2 for lunch and €3 for dinner, that is a mind numbing result for the 3-Person Family 3 X €180 = €540 per month for only Food, that's not including drinks even or other Liberosities.

Thus, like I've said the greatest error in History of Europe, or even Planet Earth, was to stop Colonization; also look at Africa since then ruled by Geurrillas, corrupted Monarchy's and other rich Family's who hold onto their Power and Energy while everyone else dies in the millions on a regular basis. Reinstating Colonization to start, stimulate and promote International Trade and Tourism with the West does not mean what it used to be by primarily Internet. We could even help build and develop local communities to upgrade and update themselves to our levels. We do the investment by a higher Currency, they make Money from their Project's and Businesses and we turn a Profit from such. Now, of course, we think that it also has potential for abuse as we have seen in various bad conditions still in Indian Textile Factory's such as the ones which burned down and the pesticide issues killing 100's of thousands . . .

As I just stated like Better Business Bureau, here in NL there is Radar, if there are not 3rd Party governing bodies who cannot be corrupted, paid off or threatened then such a plan is already doomed to fail for Greed across such conditions everywhere on Planet Earth is rampant also at beginning of 21st Century.

'A superior Economy is one in which there is maximal payback from Investment' just to purposely play on the word and quote. It has a bit of a double meaning for will we go the same way as Japan who once had the World Base Currency: The way it's going with America we could in Europe and the Euro very well become such . . . This is a great danger for only Europe and UK would benefit as the rest of Planet Earth is thrown into the worst Inflation Level's and WWIV is triggered.

And the sad thing is, each and every Country except a handful are in the red, or is this deep blood red?

To make Money efficiency in operation is the absolute mandate. This is the biggest problem in making Money. It is amazing how unorganized many are and how they can even survive which as we see now by ICT Statistic's, the last one having just come in last Friday, 25 April, 2014, that the NL Government has been losing €5 Billion on failed IT Project's, I don't even want to know for how long already considering how we have been creatively and constructively busy with Cheap and Free IT Project's for over a decade ourselves, a miserable 6% success rate only, while their Top Level Criminal's put all the Money in their own pockets only and burn everyone else like those damned Banker's who ruined Greece too . . .

Efficiency reduces costs, thereby the product can be sold for less. If one company is less costly than another, it will put the more costly company out of business. This is the longstanding Fair Competition argument to reduce prices and increase quality. However, this is also failing since now they all just make agreements with each other. Well, watch out, if you scratch each other's backs to much you'll bleed out your spines eventually. There are also plenty of Small Fish in the Ocean of Internet now . . .

People's only worry is Money; I've never agreed with their Top 03 Stress Factor's in Life, all of them require Money. Now we also have fake virtual not already existing Credit and Virtual Credit on Internet problems to add to the daily Chronic Migraine and in their cases as we hear stories and International New's Item's of how Family's in America are being wiped out and completely and dying next to all the unending incurable SAD's.

Here are the ways to reduce costs as supported by many instances, since I'm practically quoting each of them with this Cost Efficiency List:

1. People—Payroll is usually the most costly expenditure in any business and everyone makes the same error to cut too much here for next to unhappy workers, strikes and walking aways you cause demoralization, destabilization and degeneration in not just your company but the entire Economy. Next to trust factors, image and reputation you make Enemy's since it's not just a couple people effected by Bad Times but millions who can no longer pay for their monthly costs and after 6.613 Credit Card's just declare Bankruptcy and the whole Viscious Cycle repeats itself again . . . A Better Welfare System can also help against their your Boss Is God Already Complexes.

2. Service—If you provide insufficient services in multiple departments then as we have always said in Canada too, 'If you can't deliver my product in good quality in 24-48 hours then we just step over to the Competition.'

3. Product—The less you spend on the product, the more you can mark it up, thereby ousting the Competition and making a bundle yourself. You can still sell it for a low cost but then all their Greed takes over again, and yes Poverty sucks unending depending on your Welfare System. So many make this error too, since as I just stated the End Of Colonization of Colony Planet's, if you do not get your product from a cheaper source then what the Hell do you expect? And the numbers in South America are atrocious too, they do 1 in 12 high risk gambling investments like everyone else which is 8.3%.

4. Technology—Now, more tools and machines, one time purchases, plus repairs, can replace hours of People work, discount when bought in bulk. Tools since 8000 BCE with the first Planting Stick allow for greater and more stable ease of, faster, operation, important tasks are not left in imperfect Human hands or bungling insufficinetly untraiend clutzes next to large Farming Machines, 'Tjo well, I can always try to sell a toe for an arm . . . ' 'No! They are not the same size and of different value!' Holy Gruesome, Spy Kill!

5. Education—Thus as you can see by the purposefully placed misspelling, hopefully you do see it, if you don't even know how to spell, read, speak and listen goodily toally to your own Language or others then you might as well already hang yourselves by your own ties and if you have enough rope then we'll smoke your dope at your blast to the past blown after party . . . This one according to World Statistic's is actually the worst of them all since the vast

majority of systems are still stuck in Colonial Education System's. These have no relevancy anymore with Internet and the growing strength of i.e. MOOC.

6. Research And Development—According to a Scientific American article from 2013 the R&D Statistic's for all Country's on Planet Earth is completely atrocious, the best being like about 12% only . . .

I'll let you Fill In The GAP's for the next one but we don't see a 7 coming any time soon.

'Woohoo, finally Internet, so, great, yeeha, I finally get to become a Millionaire Overnight On Internet Complex! Everything is working fine, we know everything we need to, ok, I'm going to click on the Enter Button now!' Click, KABOOM!

As an MCSE'er and Webdeveloper, next to being primarily a world-wide published Author, despite my half meter RSI extending from my right shoulder blade all the way up to my top right neck vertebrae, I don't know how many thousands of times I've seen this happen and/or do myself, I think I even blew up XP Homey Boy Outdate Shit over 200+ times myself threatening to Format it quite often which pisses it off, and now XPPro Has Gone Kaboom!, many like and use this expression which someone made up in The Free Show which has now all gone to 2nd Part of Spy Kill's, and now it's on a daily basis by all kinds of Weak Outdate Software being hit by Chink Bitch Hole Hacker's who must have memorized entire Win7 already and are doing Double Blind-Folded Back Flip's with Cyber Light's in what is already turning into genuine Star War's . . .

Well, such is true as demonstratd by the Money problems everyone is having, so don't expect Internet to be anything at this time but a 16-18 year old on Sex, Drug's and R&R.

You are in debt, the store you shop at is in debt, the entertainment company is in debt, the provider is suffering massive step over and Competition Threat's, your company is in debt, your country is in debt, your Corporation is not even a Liberal or Socio-Liberal but is a tyrant who gives only a handful or one Individual the Power Of Signature of GOD and in their cases blatantly SATAN. It's proven with all the Human Sacrifices to date.

No one can spend as much as they need and want resulting in 'we are all just still and only trying to survive'. This, due to our 0 Interaction Relationship with everyone who I knew in Canada and they don't even have Social Media Profiles or photos on Internet trying to act like Ghost's On Internet which to me is 0 Hit's and you do IT Noob Suck, is about the only thing I would now quote my own Father on, who looks like Dark Sean Connery these days, not that I blame him for being an Old School Pict, that's actually a Scottish Compliment where he comes from, for it has now been proven to be true.

All of this is the natural development process of History Of Humanity which is also now being hit by multiple simultaneous Armaggeddon's, but is quite unnecessary.

Here is the answer: The way to solve the problems of Money is through Methodology, thus a valid System Of Operation. If you look at your own systems now in 2014 you see that they have ALL failed, 'You have to look in your own rear view mirrors and mirrors, you each have a Shadow Warrior on you as you have gone from A → Z.'

With organization there can be ease of flow, however if you poorly define, decide, determine, departmentalize, delegate and DOD then you are already caught in a Prism Of Your Own Design and will fail at about letter L before you reach any MMM . . .

Ease of organization = Ease of cash flow.

Easy Organization = Easy Cash Flow!

And not what everyone is doing to date: Edge-On Galaxy NGC 4013.

Now wouldn't such be nice, if is was easy to make money. I, myself, and me = 3.61322 in this department my whole Life have never heard of it, I've actually according Income Level Statistic's in practically any Modern Western Civilization Country's KABOOM! have always lived at the Poverty Level which from my perspective actually insults NL Welfare System but is smack on the nose across the Atlantic, also my whole Life I've done Cheap and Free Project's except for twice, first at HandsOn in Amsterdam but the 4 hours per day of Public Transport was too much and then at DTO who is now DICTU, I never really walked away, and definitely sprint away from that shit, I just went on to the next one, if you don't mind me to do a Final Retort to their highly offensive, rude, insulting, harassing and abusive NL Media and others with their daily shit crap content about me and/or to me, 'I don't need to tolerate jack shit at my own HQ and actually I don't need to tolerate anything whatsoever from them, I am not your Walking Talking Punching Bag, or would you like me to use your whole system in return as one?'

'No, for the 10 thousandth time, you do jack shit for me, there is no other medium but Internet for us, and NL does not account for more than .0125% of my entire Planetary Earth Hit Base, don't forget your places in the larger picture, rather than your 4-Meter Wall Mentality's, your Blow Outdate Opening Hour's and your CCC Complex, you are also a Microscopic Dot on the Global Map, have only a Handful Of Friend's Left, and the only thing inflated around here are your Ego Complexes, you're better off me an NL'er from birth born and raised in Canada to tell you this and Insult You All In Return, or get your fucking Smoke Your Own Shit and Shove Your Own Ellende up your butskies everyfuckin'minute.com in return and like i.e. if you don't drop that completely stupid retarded Paid Off Issue in your Benzine Enslavement since your Car Is An Alchoholic of let's try and irreversibly Label And Image Kyle Again the hell out my system, I don't need or want to hear and I will not only sue you again and again but I will attack each of you and your blow hurl suck out Analog Propaganda State Run Medias in return.

They have only been a Life Wrecking Ball to me and my Cheap and Free Project's to date in multiple Illegal Competition instances with all their Spy Stalking and

Victimization and then like those they apparently criticize are the Worst Heresy Artist's imaginable themselves . . .

This is not only Renting And Raving but is a Primary Cause through Total Invasion Of Privacy, Security Hacking and Business Sabotage of being each of our and your demizes to date across the board: Look again at Stock Market's between West and East over the last decade, and longer, if you still disbelieve . . .

In fact, if you are not organized, not applying a System Of Operation, not just Improve Economic's, Improv Politic's and now they want to do Improv Warfare, that's not even ordered regiments, then you are not going to make any money at all, and as we see now in 2014 after a lot of PAT our DAT's that everyone is losing Bucket's Of Money except a handful of Multinational Corporation's and their Umbrella Corporation's.

A correct System Of Operation will give you efficiency and it will give you Money at all levels from Individual, Corporation to Government, the latter actually being the worst guilty party in this matter to date with all their cushy €400,000.00 Salary's per year and all the elaborate spending for even just dinners.

A correct System Of Operation is the correct application of 6 Law's Of Economy as stated above in this chapter, they are not really mine and are quoted from multiple sources into my Cost Efficiency List.

And most definitely not the elimination of each which everyone is presently doing. And not the cutting back of each which only puts Money into the pockets of a very few. I'm not really Anti-Republican, except maybe their worst Right Wing Harvard Camp who is completely incompatible with my York and Cambridge Camp's, and your outdate HTML Interface still blows sucks hurls losing even Menu Persistance, do they only generate Wannabe Dictator's and Terrorist's with their own Extreme Opposition that only Insult's And Provokes the Enemy's to such an extent that they Trigger The Conflict Themselves, but what are they going to do Rule The World with Ivory Tower's with Mongol Hordes surrounding their walls and the Pillar's and Foundation's of Civilization cracking and crumbling as the World Platform wibble wobbles and finally crashes down into the Abyss off of Only One Pillar remaining, the Hell's go a lot further down than you think . . .

If you cut these things excessively your Businesses will disintegrate on all levels as those are the basic building blocks, your foundations, which without you would be in the mud still pulling a horse cart.

These 6 Law's Of Economy can create a stable and efficient economy, that's why it's also an even number to generate more stability rather than Only Hierarchy System.

It will, in addition, allow for the Future of development of History Of Humanity for next to having to Escape From Fossil Fuel Age Syndrome, we have to Terminate the Bust And Boom System and instate a Slow And Steady Gain System.

These are very Real Object's affecting millions and are very needed and wanted by many so if you are outvoted by a Majority Rules across not only your Government,

United Nation's, Board Of Director's or Stock Holder's but also by Leader's in Sector's than you cannot use a Veto for you would isolated yourself in the worst possible form and fashion and not be able to Colalite with anyone . . .

Unemployment is a sign of an inefficient system, too high unemployment is a sign of a illing SAD System which is already failing, so you are also only doing in your own Children in the Future who also have no jobs by also i.e. excessive over-population explosion and the resulting consumption of entire Planet Earth again . . .

Once Money Is Made

'People work very hard to have it easy.' I've always liked this one and also, 'People fight a lot to find Peace.'

This is a dualistic Basic Self-Contradiction, though not totally as one can then retire at 50 and live on beautiful beach resorts for the rest of one's Life, so long as they have not wrecked their health in the process, the paradisical dream. See a previous chapter here for what I think of their so-called American Dream.

Such is all fine, except there is a problem, such being the explanation as to why so many rich people are unhappy. Statistic's prove this and Celebrity's are also pissed off at the interference and/or invasion of their Lives.

Just think of it, for everyone has made this error too.

One earns $20000, has a 1 bedroom house.

One earns $40000, has a 2 bedroom house.

One earns $60000, has a 3 bedroom house.

Do you have any Money left for yourself? No! Are you working any easier? No! You're now just spending more and more and not gaining anything in some Mad Manic Rush to increase your Lifestyle Level, what a bunch of crap, then Go-T shows up and becomes your Neighbor with all their blasting music too, 'Congratulations! You now live next to a Celebrity!' I have nothing against him but for some it's more like their worst Lucid Nightmare coming true, or also a good example is two TV's and/or Radio's of different Ethnic Background's competing with each other in B-Motel's with No Insulation's, again.

It's a bizarre phenomenon to see people work harder and harder as they get richer and richer and then Celebrity Goes Kaboom! because of one Poeperazzi Asshole Photo, 'No, get the Hell out of our Privacy's, are you also going to zoom in on her pulsating vulva as she blows me?! Catch their own fucking Poeperazzi and/or Peeperazzi doing you know what down there, across their own profile it could be a Spy Killer or Blackmail.'

This isn't the existentialist nature of gettting rich, Existentialism supposed to be denying deeper aspects, working should be less the more Money you get.

But here is the same mistake again which everyone makes.

You make $80000, you buy a Porsche.

You make $100000, you buy another car.

Before you know it you also have 2 Wives Per Day and are working Double Shift's just to pay their Bill's across B—Stupid Violent Black Arabian Humor.

'Careful at that rate you won't have any Spinal Fluid left . . . '

And you get to keep paying out more and more because you are buying more and more and to keep up your Level Of Consumption up you have to keep working more and more until you shop 'til you drop, drop over dead and PacMan MUNCH the whole Planet again, 'Sry this Planet has been completely depleted about a month ago, you'll have to reroute on to the next Space Tourism Planet . . . '

And according to my own Uncle David in UK who is a Workaholic or is it Drop Dead From Hi!-Risk Heart Attack Alert this doesn't stop at retirement since they have an addiction and can't stop . . . 'What the hell is the point of working that hard your whole life if you don't retire in riches at your own resort??'

Now get this as to another reason next to Relationship Rules, Marriage Law's, North EU Inflation Rates, 2nd Amazon and Cheap and Free Project's as to why I want to not just move out of NL but out of EU completely, I could go back to Vancouver, Canada and also draw Unending Social Welfare there, I only need my PC and an Internet Connection, hang out like my good ole Hippy Day's next to UBC, also enjoy now and zen some Organic Dynamic Grown By Full Moon Marijuana: Here in NL, and last I checked about every other EU Country now ruled by High Elves In Belgium, you cannot register your own Business on Social Welfare but if you're a 65+'er then you are allowed to start your own Business, 'Is the whole world upside down and backward?' I have complained about this from Toronto, Canada back zen when they also have this blow suck hurl Stupid Retard Outdate Issue which we are still tracing it's origins to GOD Only Know's which Colonial Century, if not millennia ago . . .

Money isn't partial to Age, except that Teenager's have the most expendable Income and little crying Johnny or Julie who then start balling, screaming and reaching 24th Octave Decibel's again in the Shopping Mall for an iPAD, 'Oh, you are 50 now? That's OK, I will take care of you, across your QALY . . . ' Holy Gruesome, Spy Kill's!

Your spending taste does not get any less. In fact, you should be able to spend more! However, the trick lies in keeping to a 1 or 2 Bedroom House while you make $100000 Netto smackaroo per year. Your Wife and Children are not guaranteed Bread Earner's.

Here is the next Critical Error which everyone made to date living it all high and mighty with your 01 Bruto Salary Effect, that after Taxes is only technically speaking from Salary to Payment in Bank Account to A → Z Shopping Sprees only 25% at best of its Original Value, 'The whole reason we have an NL or EU Deficit is because we calculated everything in Bruto Values to date . . . ' HOLY GRUESOME, SPY KILL'S!!

'Whatever you do, never exit your Micro Budget, everything on Internet is a temptation.'

'You can swipe your whole Life away in 1 Minute Special Offer, not just by a typo.'

Such would be brilliant, more like his worst Sensai of SJ Brightness.

Imagine all the things you could spend money on if you didn't exit your Micro Budget and had stayed within your Month Budget and Spending Capacity to en faillite your whole Family and Hi!-Risk Chinese Casino Investment Plan's. You could buy the Porsche outright and not even on Credit after only 3 years.

This is the next Critical Error everyone makes which is his Parody in it, 'I want it all and I want it now!' Credit did not fail to destroy, and is still doing so, the West and East. Here it's no ones fault except you know who in History Of Humanity, once again that many centuries ago, if not millennia . . .

'Money is a result of hard work and not throwing Money into the wind like it grows on trees . . . ' That's combining a couple quotes by paraphrasing allowance.

It is for many, 'Don't forget your Millionaire's Market', more a question of want. The point is, to be rich, you don't increase your Lifestyle in all ways with your Bruto Income. Otherwise, you make no Money and will have nothing left to retire on, or take 50 years to save enough to retire on. See Moore's Law or Law Of Moore, to use my Name Convention on it does not copyright violate anything.

Now we have next Critical Error on Internet, 'When they can patent it faster than we can type it in, across Leaky Weaky BG's, then it's already End Of World's.'

Like I said though in my premise, I'm not a Dark Pessimist, always a Light Optimist . . .

Self-Sufficiency

What are we working for?

And this is not asked in an alchohol haze of mass depression.

Things are not all hopeless, even not for those in a State Of War: You still have 2 Option's: 1. Stand and Fight 2. Walk away from such violence and death.

Whoever argues things are completely hopeless is an idiot.

Those who argue for improvement are Angel's. Those who argue for demise are Demon's.

What are we working for as mere Mortal's? I say we work for the embetterment of Society and not their unending Pessimism.

More than a shallow grave, such is for sure, for at least I will die in Battle to save our Nation which is being swallowed whole again . . . This does not mean per se pick up a gun and shoot the Neighbor, that is what the Front Line is for, but to provide ourselves Near-Infinite Defenses.

What is possible?

We are working for Freedom and Peace, but more often we have to Fight For Freedom.

Strange bond, isn't it?

What are we working for? We are working for Freedom, more liberalization's and liberosities through liberations. Aren't we? Maybe we should reconsider what Freedom and Peace really is . . .

Is Freedom being up to your eyeballs in debt with a car, microwave, mortgage, with at least two kids hoping to get educated, not to mention a job in a Country with a debt bigger than the collective debts of its occupants, knowing you ain't gonna be out of debt for the rest of your life, plus about 10 years?

Sure it is. You want to do it, you wanted to do it and you are doing it.

You want, I want, we want all the luxuries this system has to offer, like free medical treatment for our nervous degeneration, 'Free drugs man . . .'

We want fast food.

We want food shipped from California.

We want everything!

Well, how about what we need . . . how about what they need . . . how about what I need?

What about Freedom and Peace again?

If you don't have what you need, you don't have Freedom, in fact you die!

Do you have what you needs?

Do you know how to grow your own foods?

Do you know how to make your own clothes?

Do you know how to build your own houses?

Do you know how to heal your own sicknesses?

Do you know how to have freedoms?

If you, or your Country, is capable of all of these, without anyone else, then you are self-sufficent, otherwise you're still dependent on local and remote trades.

If a Society knows how to get these without Trade then it is self-sufficent, however as stated in the previous chapter here, how can there be Commerce with One Global Currency, with no Currency differences . . .

Do you like to be dependant on others?

If you are self-sufficient, then you have the ultimate freedom, however how can even a Colony Planet be fully self-sustaining . . .

Just consider what it would be like, if each person was educated well, they knew how to grow food, make clothes, build houses, heal their sicknesses, play a musical instrument, be capable of anything and everything, however such is in violation of Specialty's, once again, 'Do I need to learn only one branch or the whole Tree to become Enlightened and Immortal upon the Celestial Sky's?'.

Well, if we want each and everything then we have to be capable of each and everything within the Civilization itself, however due to limited Resources and Territory's we cannot possibly achieve Independence Day Ad Infinitum since even with my proposes Animal Appartment's, Plant Appartment's and Luxury Appartment's the Alien's will eventually attack from another Space Sector, just like various Species and Races have attacked each other here on Planet Earth already . . .

Impossible? Pointless?

It is easily accessible to grow your own food in any climate with Animal and Plant Appartment's. Just grow indoors in colder climates. Do this in your attic, build a new floor, use all the extra space. You'll have a lovely lively house in no time.

Consider all the money you'll save, how much cheaper food would be in your Country, just like the Energy Network you put back into the system.

If you know how, yet unfortunately you keep lying, cheating, stealing and killing from each other when there is no reason for such when building vertically.

Don't forget Science exists. Every problem has a best method. Science has created many tools for us. It is now possible to do things with one machine what used to take many people. There are now also Hyper Modern Prototypes of which very few have been fully tested yet already implemented and utilized.

Science and Technology can devise devices to help you grow your own food and about anything else; with the new 3D Printing the sky is the limit.

Science could you give you a personal seed planting machine, could allow your country to grow all its own food and grant you all your Luxury Item's through import and export of such basic and advanced materials and methods to do so eventually leading to genuine self-autonomous States and Country's instead of conducting War upon each other.

You could learn Science and Technology and apply it to your buildings.

Consider all the Money you and your Country would save, not to mention lead to the prophecized tentupling of the population of Human Species and colonization of the stars despite all who die in between.

If you know how; you don't know how for with all of the Information on Internet you still don't know how to apply it correctly with the most going into the wallets of the few.

You could know exactly how to pour cement, make cement, get the ingredients for cement, steel, gold, silver, glass and cloth. It could be one Planet Earth with everyone helping each other to build rather than ripping each other off. Consider how expensive it has become to build one single building, yet we all have the same goals for our own Children to the Star's, allowing for the necessary Competition.

One self-sufficient Planet Earth, not warring neighbours, not capitalistic greed, not commi control and not fascist hate.

All other entertaining activities can fit beautifully into this. And who says we are denying their demises now in their own Armaggeddon's across their left, center, right extremes in 2014 CE? How many entertaining activities do you get to with your 40+ hour week?

Things could be far more efficient in the application of Import and Export.

Would you rather work a 40+ hour week in a boring repetitive stuffy office in a big building with no opening windows, so you can go home and watch TV for 65 years, then die after your own have all left you for being a boring outdate?

Or would you rather, or your Country, apply yourself with you own hands and your own brain to your own design of your own house and your own clothes, and your own food, and heal yourself?

Consider how expensive Health Cares have become with their blundering idiots who wouldn't know how to shove which Ill Pill in which direction, always ignoring causes and premises ever since their 2.5 millennia old regime, hand in hand with their 2 millennia regime and no better with their 1.5 millennia old regime; since when does Religion and Politic's decide your per per physiological Diet and Fitness?? Absurd!

It all rests on Science, Technology and Education, 'Only when you find a way yourself will you find the will and only when you find your own will will you find the way . . .'

Educate yourself! There are no excuses left with Internet. Export and Import yourselves, what do you think they did to date? Only then will you recover from National Debt's, Bankruptcy's and other Debt's, only then in the Future after many decades will we reach some form of self-sustainability through self-sufficiency.

Since we have all been brought up in the Classical European Idealistic Education, with primarily Classical Empiricism ruling, only now do we step towards Quantum Sciences in beginning of 21st Century, we don't know anymore what the good old peasants knew, what the Celts knew, what every Society in History knew, allowing for their suppressions and repressions through many War's: How to supply the basics which a 3-4 Person Family Unit need and want; this is ridiculous these days in North EU Rip Off Zones.

Consider, with bust and boom, how poorly balanced the economies are, now . . . still . . . all increasing in debt only . . . instead of slow and steady gain . . . with a 4 Person Family Unit we still only maintain our numbers . . . not even increasing . . . just another 'sorry it is too late now' Backwater Colony Planet.

We are a fortress with no foundation and we have soil erosion.

We need practical and better Human development and sustainable Education. And we had better get it soon. We need sustainable development on a global scale. We need to work in Balance With Nature, not rip strip rape plunder pillage deprive and deplete it.

We can't have huge fields of crops sprayed with pesticides, picked unripe, nutritionally deficient, and all the soil slipping away. The Rainforests are being cut down at the size of Greece per year as according to Greenpeace statistics. The land is wasted to pig, cattle and sheep, all their manure, is poisoned and what are the wasted percentage of our bio-organic crops left? To date before Industrial Age each and every last crop was bio-organic; some of the worst pesticides were introduced after WWII from the Nazis . . .

Mother Earth is being ripped and raped. We can't pollute.

We need to learn how to live in Harmony With Nature, with a future 20 Billion population, with self-sufficiency and sustainable development, to populate the Solar System and further, to save ourselves and prosper, for it's already now in 2014 a half-barren Planet with no M-Class Planet in sight . . . Can anyone deny the possible catastrophe which will occur with such a population and present mentality and methodologies? We can barely hold 6.6 Bernard's together which is the number it's definitely suffering from . . .

We need to face Reality. We need to learn through Observation and Experience. We must not remain in States Of Denial's, Ostrich Effect's and Dump Sand Over It All Again.

The Education system sucks, stucks in Colonial Ages with Locke, and locks and loaded on it as being the primary cause of our demizes in Modern Western Civilization's.

All the tools and information are here. We now need to apply it correctly.

There are too many greedy Individual's, Government's and Corporation's who have lied, manipulated and put down others to gain their own. In the process, they have destroyed the world; how will we ever recover at present over-explosion of population of consumption rate from a half-barren Planet if we don't start Animal, Plant and Luxuy Appartment's pronto so we can get Resources and Territory's from other Planet's, and finally exit Fossil Fuel Age by stepping over to all Types Of Hybrid's . . .

However, it's the uneducated people who are the victims. It's the lack of education which even allows for these sycophants on their victims to do anything. They do not allow for independance, self-sufficiency or sustainable development, they only want to satisfy their own Power and Energy hungry greed complexes.

Please educate yourself! Observation and Experience. Philosophy. Science. Education.

And keep in mind if you, or your Country, know how to supply all the needs and wants by yourself, youwill never have to worry, never go to War, nor will you ever be a victim, you would have Freedom and Peace through self-sufficiency, sustainable development and Near-Infinite Defenses.

Discretion And Integrity

1. Integrity

Work is exactly what becoming Immortal and further involves; you each thought it was some mystical, magical and/or mythical secret, no you simply work your way up through the Rank's and echelons of Planes Of Existence through multiple Lifetimes. Next to earning a living, it is the only way to pre-occupy yourself on a daily basis and become happier, work makes you happy, otherwise you sit around doing nothing all day.

And yet, very bizarrely, a cool quirk of the Law, All Things To Balance, we have a serious desire to do absolutely nothing, to be in a continual existence of total effortlessness. We want to stay in bed. Bed is the greatest place in the worlds, where one can be in a Dream State or have Lucid Dream's, free from the miseries of the world and all your blow suck hurl binding Relational Argumentation's.

We create a vast quantity of convenience through Science and Technology to reduce our expenditure of Power and Energy. We become wealthy so we do not have to work: 'People work very hard to have it easy.'

This underlying pervasive lack of effort is the result of us all being witness to those who do not have to raise a single finger to survive, such as lucky Little Johnny and/or Julie and his near-unlimited Inheritance. These include children of wealthy families, vast inequalities throughout the whole system, the actual lack of free oppurtunity for immigrants, liars, cheaters, stealers and hackers of the system and others who benefit from windfalls coming from Lady Luck, like lottery winners. There are so many imbalances in the system by such that you have to ask yourself if there really is any system at all but only Money deciding it all again.

The rest of us have to slave away the majority of our lives, 12+ hours per day Full-Time, for an average income salary, or even less . . . Combined with all the fallacies of the American Dream, plus the other necessary daily activities, you have to ask yourself if you even have 1 hour left for yourself in a day. Freedom is, therefore, always relative, 'You are bound by your freedoms.' However, you made unconsciously, half-consciously or consciously your own decisions and your own Power Of Signatures back then, now trying to blame others and the world that it all went wrong again, 'Hah hah hah, did you each right back in, always get even and always stay angry . . . '

The above challenges our inherent beliefs and faith in Equality.

The Type Of Work and Degree Of Work is what sets one apart from another. In Modern

Society, and even primitive ones, this determines your entire existence. The rest, such as your ideals, inspirations and dreams are not significant in the eyes of

Society, only your net worth and your net potential worth counts; only a few lucky Artist's get to become rich and famous in the unending pecking order, if you even want to become such with all the excessive negative and/or positive attention it brings. As I stated in a previous chapter here and in my other works in New World Order you are no more than a Planck Number of 9 digits for your Unique Identity stamped as bar codes into your arm like in Brave New World: I am 0.123456789 the God Personage who finally did not fail to emerge, arise and take control of all the other Multiple Personality Disorder's, 'With 313 profiles on Internet and counting what the Hell do you expect?'

One has to become One With Nature for development of one's Evolution. Pleasure is what you are mostly looking and working for, yet each Pleasure has a Pain.

The tedious persistence of rest and laziness is due to ambiguous convictions about gainning physical pleasure. If we know we can be guaranteed pleasure from work, we will work hard and happily. The reality of working in the Modern Work Force since Industrial Revolution, especially in the 21st Century with Information Technology, is completely something else. The daily forced bump and grind amounts more to a continual pain experience just to keep your neck above water, the rising costs of Inflation make each netto $ and € worth less as each of our individual and collective debts rise; This makes it 2-Bit Obvious that the Power's and Energy's that be only want to enslave us for decades and Lifetimes. We will not tolerate this for it is in violation of our constitutional Right's Of Freedom's, Liberation's, Liberosities and Peaces. Obviously, if you swiped your card or wrist like such or broke the Law's of Country then you have only enslaved yourself.

And it is a sad and violent History Of Humanity of the litany of Power and Energy mongers. 98% of the world population lives in poverty or low to average Income and 2%, through primarily rich and corrupted Individual's, Government's and Corporation's own the world. It's only getting worse at the beginning of the 21st Century as Multinational's buy up everything in known existence. You can make money with money, the rich getter richer and the poor keep getting poorer . . . There are now already 14 Umbrella Government's or Corporation's in 2014 CE which will soon be consolidated into 7 and then 6, and then only 2 eventually, as the two opposed tyrannical Titan's battle it out.

You can work at being a total jerk. You can work to become Rich and/or Famous. The hardest way to work is to gain Integrity. This involves discovering Truth and then actually putting it into practice, regardless of what the price to pay is your Integrity will always be awarded for it is not only governed by highly subjective Multimillionaire Puppet's but by Law's Of Nature and Universe, all things to Balance through not only Karma but Quantum Physic's as even Classical Scientist's are being forced to admit.

Nature does not give anything to anyone. There are, through the hard work of Evolution, no Gift's. No Genius ever made it without working everyday. Something does not come from Nothing.

Each step of the way has to be earned. This does not per se mean Blood, Sweat and Tears by killing the Enemy and ripping off their Resources and Territory under the guise of all kinds of Lies And Excuses.

The work of progress has more to do with Pain then it does Pleasure, since Pleasure is an experience of rest which you do in your free time or reap the fruits of your Labor, however now since all their Capital it is no longer a result of real work but fake stocks.

To work to become One With Nature, in the long run, is increasing your Integrity through honest work and if you are not awarded in this one then you will be in the next one. Yet, everyone looks at only short-term and personal gain leaving none for their children, and leaving very little left of an already half-barren Planet Earth also wrought now by War everywhere which is no longer swords and shields through great forests and plains.

Integrity is Knowledge plus Good Will put into practice together. Instead of continual persistence of greed and corruption, not to mention the mass killing of the innocent to accomplish such goals which are nothing but the temptations of Lucifer and/or Satan, and other malovelent Demon's and Spirit's, who keep promising you unlimited Power And Energy and oh all the Immortality across your very Not-Immortal Host Body, they want your Soul's and Spirit's and will enslave you for centuries and millennia.

Integrity is Truth in all things at all times, your Trust Relationship's in personal and business matters all rely on this, and like anyone else they took their Not Unholy Revenge upon you for your dishonesty and betrayal. Integrity is the hardest work.

It is also, through Karma and Evolution, the most rewarding. You cannot take all your gold, silver and copper coins with you to the grave, not to mention the cute little boat sinks before you get across the River Styx . . . Big swimming, jumping and flying fish eat you before you ever get to the other side for it is a Hell of a lot wider than you think. A huge mountain is to the north, a great waterfall is to the south and behind you is fuzzy reality, your previous life with your dead Host Body which you can never return to . . . Only by paying the boatman one 05 copper coin will he grant you save passage to the underworld where you must walk through onto your next lifetime . . . This is Greek Philosophy, Religion and Legend and I equally do not tolerate my Scottish Celtic Habit's And Tradition's being insulted by uneducated brainwashed Noob's like yourselves.

Integrity successfully continued at all times, Nature returns to you continual Pleasure. Think, for example, of the good feeling you have each day knowing you saved some African Child and gave him/her an Education, not only over-consumed in the West. These days people think only about themselves and their own children. Some though are so bad, they literally throw their kid off of the tree branch, again, to make Money, your whole blow suck hurl lives, for the pursuit of their lies of the American Dream, revolves around only making Money, those who do not succeed are perceived as Sick, Ill, Weak, Mentally Disordered, Psychologically Impaired and/or Criminal, outcasts of Society, Abnormal Miscreants, Mutant's, Evil and/or Homeless. Yet Society itself due to such mentioned errors, imbalances, strife, prejudice, greed, corruption, inequalities, warfares and tyrany to date in entire History Of Humanity has actually caused such.

With the correct application of Integrity to all daily activities you and the Country and Planet Earth will make more progress into the future: As a Great Indian Chief once said, 'You do not own the sky, the earth, the seas, the plants and animals under God, you only temporarily lease it . . . ' So we better take care to be better Guardian's of it with the necessity of Laser Military and Near-Infinite Defenses with self-sustainable development of Evolution of Humanity. Otherwise, we will mostly likely cause self-destruction through various factors such as environmental deterioration, world poverty, and unending warfare. Of course, based on the size of China alone now, we will survive as a Species but what, truly, is the quality of your existence?

2. Discretion

Living a life of Discretion is ultimately more fulfilling than a life of indulgence and consumerism. Lasting happiness, although it first seems contradictory, is obtained through a life of dynamic discretion.

The first argument against being discretionary, materialism, is the easiest refuted. The acquisition of material goods is a trap. Concentrating one's Power and Energy on material possessions brings no lasting fulfillment; the consumer always wants more. The material wealth grows boring and tedious after awhile. The new car is always outranked by a later model. This is the way of material things. They always change. In the end you just have the same rehashing of the same theme only put into a better cover with better graphics, such as 3D Film's and Games, still just another Hybrid Clone Complex with the very same Topic's and Themes, Heroes and Villain's, God's and Goddesses, GOD Complexes and all the Archetypes of the Unconsciousness as depicted by Tolkien, Jung, Freud, Grimm Brothers, Roman Mythology, Greek Mythology, Celtic Mythology, Egyptian Mythology, Chinese Mythology and Mayan Mythology for the correct order spanning back into entire History Of Humanity, though there are also others.

The same plots and characters, simply put in a different form, would that be because, 'Lessons not learned in the past are doomed to be repeated again and again and again'.

Such is the way the spritz and excitement, like your first week on Internet, dwindles into a dull blob, no longer sparking interest, no longer granting fulfillment and causing more problems than it solves, with unending errors, glitches, bugs, hacks and more costs.

The way to approach material goods, rather than wasting every last minute of every last day in the pursuit of material acquisition, is by using them as practical tools, a means to an end, which give benefits and hindrances to our lives. Truly, not denying it, Science and Technology has and will provide us with plenty more Leisure Time. Nicht Ontheil Pythagoras! However, can you pay the price tag coming with it? Can you then pay the next price tag coming with the next generation? Can you pay for the repairs on a device which has broken down? Do you even have any Insurance again for your smartphone, still all Stupid Devices if you ask me, since it costs an extra +/—€10 per month? Is there any Hack Insurance, again, or as long as you're a Goody Twoo Blue Boy or Pink Girl then it will work this time at midnight while listening to your media player . . .

And then in the beginning of the 21st Century, the out-of-control Credit System has lead to a world-wide Economic Crisis. Now to get their Money back, again, they as are so-called Near-Enlightened Leader's are risking and triggering WW04. We only hope again and again it's a temporary bust and boom, a short recession, but everytime it's the same Poly-Cracker On Crack Cocaine Bullshit Politician's and Busyness Leader's, and how do you in GOD's name recover from a stupendously huge national debt? NEVER! I keep saying you should just Tabla Rasa the National Debt and put in a new referendum, but that would obviously not sit well with their Multinational Bank's who are the real bloodsucking Alien Vampire Demon's with their Black Shadow Vice Grip's Of Death.

If the benefits outweigh the hindrances then fine, use them, but do not be so foolish as to become emotionally attached to your new microwave, like it's your new lover, or something, before you know it after Made In China and Made In India and Made In Taiwan it'll only be good to stick your you know what in.

Internet as an Information Exchange is plima, but also think of the billion dollars of damage it is causing to businesses and sectors everywhere. Will I be replaced by a Cyborg in the future? Or will I be only an iCyborg? What happened to my PentUp Cyborg who talks back only when I say so?

And then you know who has a much better XYZ-7700 2025 Deluxe Model Cyborg Pet.

How much is this Hyper Modern Future going to cost?! And for who only?

Discretion is simple, though hard to apply. It avoids the complexities of wasting your time, your whole life, on acquiring money to acquire goods, to acquire temporary happiness, to then buy more happiness with upgraded models, before you know it we'll have to also pay for updates to our OS's and Softwares, oh wait, oops already . . .

Discretion avoids becoming dependant on a lifestyle of wealth and material acquisition, which is why you slave away 12+ hours everyday. What kind of life is of sugary cakes, mansions and 'security'? I could crawl in through your open window and threw bricks, paint and paper at ANWB here in The Hague 6-8 times in a row as a Greenpeace Activist from Nov 2012 to date May 2014 cause they have two god awful dismembered 50 year old trees in front of their building, since they have no Security Cameras, no Car Alarm's, no Alarm System, no Security Agent's, and hide behind coincidentally happening to own that 2-Bit Square Meter land property and why not rip all of the hubcaps off of their dark blue cheap Ford FBI cars too, I had to walk on purpose through a red light in front of a cruising searching Police car to get myself caught on purpose and admit it to them as also an Agent of Not Injustice Incorporated, I mean what is wrong with you all, get some scanners even . . . It is one of degenerative sicknesses, diseases of consumption, undisguised selfishness and weakness of character through total dependancy; thus this being a perfect example of their enslavement of each of you through the mandatory costs of Benzine, Oil, Gas, Road Taxes, Fines, Driver Licenses, Car Leases, Car Check's and Car Ownership's in combination with their Vice Grip on Full-Time and Part-Time Job's stating you must have a car to function in Society. All those who do not comply with also their brainwashing Advertisement Campaign's, Propaganda and State Run Medias from the all-powerful Arab Sjiek's are hunted, persecuted, locked up, wiped out and in worst instances also killed, murdered, assassinated and/or destroyed depending on Left, Middle, Center, Right-Wing Politic's.

Who is more vibrant, the Native Indian or Queen Elizabeth? A life of wealth if too far removed from Nature to allow the human being any lasting happiness results in the new Hyper Modern Luxury Appartment's like Dubai and other where you never have to go outside anymore, don't open your doors and windows, it's an evil scary dangerous world out there, don't walk or bike outside for they'll do you in, and whatever you do never go into the dirty public transports systems cause they're infectious degenerates. If you read just a couple stories about rich families, you then see it is not money which brings happiness, you see they are the worst scared individuals in Society, all for their Money.

Discretionary behaviour brings more ease into your life. There is less worry about being fulfilled, less monetary investment in your happiness, less dependancy on material acquisition and as a result you are fulfilled! Fulfilled with such things as mental, emotional and spiritual development, the soft warm spot in your chest, the lack of worry, the fruits of Knowledge. You have your needs and wants; so to practically end the argument right here, how much does a Human really need or want, or are we doomed by some innate instinct of more and more material acquisition to destroy ourselves . . .

What type of Society produces stresses, diseases, unhappinesses and wars? A consumeristic Society, an unenlightened Society with no other goal except the fulfilling of its ONLY its own National Identity, its unending increase in needs and wants and the filling of its ONLY its own treasure chests. In those previous centuries and millennias up to even 20th Century such needs and wants, such sicknesses and diseases, leading to more and more over-consumption, did not even exist, we're now all high in one big Halluci-Nation which is about to turn into a Lucid Nightmare.

Socrates figured it out 2400 years ago in a conversation with Glaucon:

Socrates: 'And there will be animals of many other kinds, if people eat them?'

Glaucon: 'Certainly.'

Socrates: 'And living in this way we shall have much greater need of physicians than before?'

Glaucon: 'Much greater.'

Socrates: 'And the country which was enough to support the original inhabitants will be too small now, and not enough?'

Glaucon: 'Then a slice of our neighbours land will be wanted by us for pasture and tillage, and they will want a slice of ours, if, like ourselves, they exceed the limit of necessity, and give themselves up to the unlimited accumulation of wealth. That, Socrates, will be inevitable.'

Socrates: 'And so we shall go to war, Glaucon, shall we not?'

(Plato, The Republic, Book II, Translated by B. Jowett, p. 233.)

Consumerism is the force responsible. Consumerism based on Credit is disastrous.

There is no Discretion in Modern Society at the beginning of the 21st Century.

It is directly related to going excessively beyond necessity and wanting. The fulfillment of need and want within certain lines and limits satisfies, nothing else. As I referenced Scientific American in my Spy Kill's 02 the West consumes 2/3 of the world's resources and has only 1/3 its population. If the present trend continues and/or other Country's like Russia, China, Africa and South America build up to our excessively consumptive lifestyles, presently at a hyper acceleration rate, then as stated previously we could quite literally have only a decade or two without the correct implementation of my proposed Animal, Plant and Luxury Appartment's. $E = MC^2$.

The 01% Rich Elite think the solution lies in Space Travel and Space Tourism to get more and more and more Resources and Territory's, but thanks to Deep Space Scan's by many parties we now see that we'll never get there in Time. Matter = Energy.

Need is Food, Drink, Shelter, Clothes and the right to do what you want as long as you harm no one. When we sit in the Sun and swim in the Water at the beach then we are happy. Modern Society has lost perspective of how many Human's there still are without these needs. The percentages from many sources on Internet are appalling, it's more like we're still in the Feudal Ages, the only difference being Big Boy's with Big Toy's.

Want is the expansion of such to Liberalization's, Liberosity's and Luxury's which with all the temptations on Internet and the far too easy capability to swipe your cards and wrists straight so smooth soft and silky has not failed to lead to downfall of Modern Western Civilization in beginning of 21st Century and with rise of East we might as well sell out.

Such a way of living brings us greater awareness of the beings around us. We recognize our own needs in them. We see them as living, breathing beings, capable of pain and loss, and we see they are not just brute animals, inhuman or heartless wandering robots. We develop a sensitivity and understanding for others through this awareness. This is what Discretion and Integrity is about.

Not that each one is ONLY a Planck Number with a net worth or net potential worth.

A generosity will stem from the integral and discretionary lifestyle which nurtures the qualities of a spiritual attitude. More people will live in harmony together, with no reason to go to War, at any level, but until people's needs and wants are not satisfied there will always be strife and prejudice and those who would take imbalanced Power and Energy into their hands from others need to be put under quarantine lest risk Planet Earth.

A life of Discretion and Integrity is not one of Asceticism. Such a word conjures up images of thin, bony, starving bodies. Discretion and Integrity are lines and limitations in levels of consumption for needs and wants through logic, reason, moralism and emotion.

These are about the only ways, allowing for some miraculous god send or windfall, that we will ever achieve self-sustainable growth into the Future without wiping out 1/3 to 2/3 of the entire population of Planet Earth as predicted by New World Order and Brave New World . . .

Let's hope not ALL the Prophecy's are true . . .

Time Does Not Exist

Reality is one large continuum. The past and future do not exist; it is only one big now in Motion.

Time is confused with cyclical Motion. There are an infinite quantity of points on one line, which then loops to form a circle, or a closed circuit. These circles can also form spheres through the Mathieu puzzle, M24, ' . . . in a 24-dimensional space. It is known that among all sphere packings in 24-dimensional space constructed by centering 24-dimensional "spheres" on the points of a lattice, a sphere packing based on the Leech lattice is the tightest.' Igor Kriz and Paul Siegel, Scientific American, July 2008.

Time Travel is not possible. Otherwise, you would have crazy Scientist's, Magician's and/or Dark Sorcerer's jumping back to every previous nanosecond to prevent WWII. As stated, described and proved in my Planes Of Existence—1st Edition—Published by AuthorHouse UK you can only make new Timelines, thus in other words send yourselves off to your own Hell Planes, Middle Planes or Heaven Planes in Near-Infinite and Infinite Umbrella Form's of Timelines.

Timelines can exist on all levels from Individual's to Galaxy's and you have an infinite quantity of double identities and universes in reality which slightly differ from each other, however your own Soul and Spirit is unique, though it may get possessed now and then, fall into error, correct itself, go into recess, develop, descend, arise, sleep, awake and finally after multiple Reincarnation's free itself from its Mortal Coil to explore the Near-Infinite Galaxy's in an Infinite Reality.

A new Timeline effectively creates an entirely new Galaxy even. You can just as easily argue if Napoleon was killed while crossing the bridge in one of his first Battles then someone else would have just taken his place and the French Revolution would have happened anyway with a mild name variation like Napelion or it would not have happened and Keizer Bismarck would be perceived as the great liberator or even worse entire Europe would be still Nazi Germany; very small key critical Event's changed the course of the War with only Normandy being the big climax of their deaths.

Also within our Universe, consisting of Matter and Energy, Particles and Waves, you can only have a Near-Infinite quantity and quality of layers of Dimension's and/or Planes all coalescing over each other, one invisible to the other, one insubstantial to the other, one immaterial to the other, apparently by Illusion and Delusion of Only Visual Spectrum. Infinity is only allowed by the great Nothingness outside of our Universe which it is expanding into, otherwise by friction alone it would burn up the entire Universe and cause another god awful Coffee Cup Effect. See previous chapters here and my other works for more detailed descriptions and explanations of these matters.

The merely subjective measurements of cm's, m's, km's, minutes and hours is a Human Invention. Timelines are confused with the vortex, vertex and vetrice paths of all their Motion's throughout the Universe, many as we see now being Spiral Motion's.

One could almost say: Time = Motion. But any Mathematician would scoff at that. Time is simply like Einstein said what we make it highly relative; if I want to travel at .98 Speed Of Light than I will do so as long as I am not limited by such Mass. This is why Shadow Energy and Shadow Teleportation and Dark Energy and Black Holes and Mini-Black Holes which do not suffer such resistance are theoretically Instantaneous Travel. Indeed, if you generated an EM Field strong enough you could punch a hole in Space Time, which I consider to be an oxymoron now: Space Motion or Motion Of Space is better and finally resolves the confusion.

One moment to an Eastern Mystic like the enlightened and immortal Buddha is an Infinity. An Alien Species does not know this concept of a minute or a kilometer having a completely different measuring system. If you walk around with no watch on, only seeing the cyclical motion with spiracle dilation and differentiation of the Sun, Moon, Planet's and other Star's then you also have no Time.

Time Does Not Exist as I also began with some of my very first works in Poetry Lore—1st Edition—Mass Energy, The Death-Life Conspiracy—Published by AuthorHouse UK because it is confused with Motion and as Human's to function we need '1 . . . ,2 . . . ,3 . . . , seconds, 4 . . . , that is what a clock is for, it gives us order, it gives us reason, structure . . . ' and we cannot measure or make anything without applying this very highly relative, yet completely Not Existent Element, like the mean does not exist in statistics, some vague virtual out in nowhere figure, which is ingrained in our psyches.

The concept of 3 dimensions, X, Y and Z, a mere mathematical chart is also not true. You can just as easily argue there are 6 dimensions, coming from a cube, or even 12 a bi-cube. If each room is an octagon than you would think there 8 or 10 dimensions. If you draw a circle, or better yet, a sphere than you have an infinite quantity of dimensions. Time, therefore, is falsely associated with the 4th Dimension, it's more like even the 12th Dimension. Nicht Ontheil Chronos!

Yet, it is none of the above, for it is all of the above: Time is Infinite, Near-Infinite and Finite for it is Motion itself. We for our purposes into the Star's simply get to apply whatever numbers we want to it just like we can throw anything into Infinite Equation's, Near-Infinite Equation's and Finite Equation's.

Time, in fact, within the infinite Universe does not exist or you also have to deny the existence of GOD for Time and Timelines inherently imply only Near-Infinite and Finite which cannot be otherwise GOD can die too and that's probably what Nietsche implied. However, there is obviously and underlying Infinite Timeline which loops on itself as still being the correct solution to the Infinite Point's On A Line by Pre-Socratic's which is my Specialty in my 2nd Year Propedeuse in Philosophy at York University, Toronto, Canada.

The Speed Of Light is not the maximum speed at which one can travel. It may be the maximum speed of Mass, as Einstein said, but there is no mention of Information or possibly Mind, Spirit, Soul having no Mass which can travel instaneously. They also purposely avoided Shadow Energy, Dark Energy, Black Holes and Mini-Blackholes due to the War, his Good Father Figure Role and honestly the world was not ready. However, now we have full blown Internet with such topics and themes dominating Internet and all Entertainment Industry's and with Planet Earth suffering horrible demises everywhere the Get To Another Planet In 266 Year's at Speed Of Light is being now antiquated if not blatantly lauged out by some for we might as well sign our Extinction.

The zero-point-field theory, just to not use my own Name Convention for once, states ' . . . no space can be truly empty since it is always full of ether, light or the base electro-magnetic matrix of our Universe . . . ' Books by Lynn McTaggart on the Topic are also good to read. Therefore, our Universe remains Finite and Infinity is allowed by the Multiverse Theory and Nothing is only external to our Universe since they purposely leave out what is 'outside' of the Universe and talk about what is 'inside' the Universe; it can only expand into Nothing or there would be a huge destructive friction at the edge of our Universe as it collides with other Universes also consisting of Matter, Energy, Light, Mind, Spirit, Soul. It could be also be full of Shadow Energy or Dark Energy which is in combination of Big Bang Theory also pulling such, however Nature abhors a Vacuum and Shadow Energy or Dark Energy also goes through Nothing. See a previous chapter here, again, for the description of the difference and confusion to date between the terms Nothing and Nothingness.

Nothingness and Nothing which could both be Infinite Null Potential which can be transfomed into Infinite Kinetic Energy allows for Infinity and Instantaneous Travel. It is therefore false to presume the Energy requirements necessary for such Space Travel would wither drain or blow up entire Planet Earth since it doesn't even use any Sources Of Energy here; picture again the analogy of dropping a pebble from the top of the Universe aimed at Toronto, Canada. Also, you ask yourself, how will Humanity ever explore our Universe if limited by Only Light Speed again?

This all suggests and provides arguments that our concepts of Time are false and based on false axioms. If it is one large Infinite Reality under GOD, or Infinite Reality = GOD, consisting of Multi-Universes then Time truly does not exist. A cyclical motion, with many curved, elliptical, spiracle and other Motion's such as the Season's does not prove existence of Time since they slightly differ each time and also change throughout centuries and millennias i.e. if we cause an Ice Age or massive global warming and if the Planet Orbit changes by meteor impact and all the other impacts and collisions to date in Evolution of Solar System's which have created the Celestial Body's.

Such is all simply Motion going around in circles and ellipses ad infinitum so as to, for example, prevent the collapse of Orbit's and maintain Universal equilibrium, to some degree until, once again, Andromeda Galaxy collides with Milky Way Galaxy. This is why I am still leaning away from Big Bang Theory to Infinite Existence Theory, I don't know if I just coined it but I definitely support it, since shouldn't everything be moving away from each other? Just kidding, Bad Joke of mine again, that is Pas Kaboom! Should they not all be accelerating away from center point of Big Bang and not two huge Galaxy's moving straight towards each other? And what is the Nihilistic Existentialist Purpose of destroying everything in the end as Black Holes swallow everything up again? Also new Galaxy's and Sun's are born all the time. See also previous chapter here for more arguments in support of such.

There is only one large Space Reality in Infinity with no such thing as Time. Time remains a purely subjective measurement to facilitate the practical application of Science, Technology, Mathematic's, Astro-Physic's, Physic's, Information Technology, Art, Archticture and practically every Sector and Specialty in existence. It is useful to build a wall or a transport system but in Infinite Reality, it does not exist, unless you punch a hole through Space and use Quantum Entanglement and other Quantum Theory's such as Quantum Transformation and Transferrance of Body, Mind, Spirit and Soul from Matter to Energy to Information to somehow aim at a distant Star in a distant Galaxy even, 'KABOOM! Oopsy, our Space Ship Hyper Drive Shadow Teleportation Engine ended up in space debris again . . .' Holy Gruesome, Spy Kill's!

'The true end of our Cannibal Galaxy's is nigh as massive forces of destruction are released, planetary bodies ripping apart, as the two massive Black Holes pull towards each other and Civilization's at their edges flee in the opposite directions, only able to save a few Elite Scientist's and Artist's, as all the rest, for the worst suspense and horror, get to watch for about a century or so before they all perish horribly . . .'

The statement of a Big Bang Theory suggesting a beginning also suggests an end. Therefore, once again, our Univese can only be Finite eventually collapsing or exploding apart. Yet, how can it then keep expanding into Infinite Nothing and/or Nothingness? Are the Multi-Universes also expanding away from each other like billiard balls? Is it not actually simply persisting, looping and maintaining itself through Law Of Conservation Of Energy which is an axiom of Science applied to Multiverses . . . Just like you are reincarnated by the same Law Of Science through Body, Mind, Spirit and Soul . . .

What is this Infinite Space consisting primarily of Nothing? No, it is, once again, as we see in between the Galaxy's primarily Nothingness. If there are an Infinite quantity and quality of Planet's, Planes and Dimension's, all overlaying each other, then Energy and Matter, Particles and Waves, Spirit's and Soul's are also Infinite. The overlapping, interaction and collision with other Lines Of Energy through everything causes indeed many people, animals, plants and things to burn, implode or explode. See previous chapters here and my other works for Lines Of Power and Lines Of Force.

To sum up the most important concepts here:

1. There are Infinite Spaces in One Big Reality. One Big Space and no Time!
2. Time and Timelines are confused with Motion and vectors, vortexes, vertexes and vertices of Motion's.
3. Our sense and use of Time is only highly relative subjective measurements.

Now unto some debunking of some more theories to date and Comedy Relief:

The concepts of Black Holes, Worm Holes and Hyperspace to date are also more full of loops and holes than a leeky siff.

A Black Hole may transfer Energy from one Universe to another suggesting therefore the denial of the mythical isolated closed system but it is no means of Space Travel. And then Hawkings, by Gravity, Mass and Density alone, you get ripped up into Pure Energy, effectively dead. So, not only can you not get back, but your are no longer alive. This is not only Hawkings Black Humor but what is the point of exploring such in Spirit Form only when you cannot interact with the environment?

The problem with a Worm Hole is not only would there be a stupefying quantity of holy tunnels through sub-space, causing huge instability, but you still need a receiver at the other end, just like Gate Technology, ' . . . leading to First Violation's of the Origin's Of Ancient's and it really is in Mount Ceyenne and they really are running our Government, and good God could we pls stop rematerializing out of nowhere into a wall, a moon, a planet and there really is a shit load of space debris out there, sir, hut, yes sir, NO SIR!'

And what medium is this twisty entrail using again?

Hyperspace even defies the Law's previously mentioned cause now you have 9.8 times the Speed Of Light in Star Trek through Space itself. And what is it going through? What are the chances you won't hit some planet, why is each one in a perfectly straight line, the course corrections necessary to avoid burning up everything in between and the spaceship itself would be even somewhat beyond Q's IQ Level = Quarky.

How then, again, within the Energy Matrix, of Space Time, Where . . . Is . . . The . . . Space . . . Ship . . . ?, are you now now Neo-Spock, do you suddenly, whoosh presto, rematerialize without turning into a squid? How even do you even aim out there?

There is, after all, nothing in our Universe which is not material, Matter and/or Energy, it is merely a question of gross to fine particles and/or gross to fine waves and/or gross to fine EM Field's and can we send her Obese Fat Ass Out There.

Where are these mythical breakthroughs in Science and Technology and Magicks apparently in all Sector's now in beginning of 21st Century, still no actual laser phaser fuzzier since Buck Roger's, or was it really actually the Nazi's and/or Ally's doing Paranormal Experiment's as to, ' . . . how the fuck, sir, are we supposed to exit Fossil Fuel Age, ala 115000 kg of fuel alone needed, with no maneuverability, in less than 100 years, you want me to do what, yes sir, I mean, NO SIR!'

Such can only be finite present Science in beginning of 21st Century having discovered the quantum particle, though not all of them, uncertainly, down to the limitless vagueness of some Scientist's to the inexplicable undescribable, nobody has any clue what they really are or do, Quark's and Lepton's which are all Fermion's and then the Boson's, like the Higgs Particle which they apparently now measured at CERN, but nothing else, and how do they work in Standard Model from Up, Charm, Top, Down, Strange, Bottom, Electron Neutrino, Muon Neutrino, Tau Neutrino, Electron, Muon, Tau, and then respectively the Electric Charge and Mass of Photon (EM), Gluon (Strong Force), Z (Weak Force), W (Weak Force) and Higgs again . . . Apparently there is also an Unknown Particle which could be Anti-Matter and/or Dark Matter . . . I'm not saying they, as Expert's are per se wrong, I'm just saying that everyone is Stabbing In The Dark since Maxwell, Einstein and Planck who have still not been debunked, except for a theoretical Speed Of Shadow, bad acronym I know, and Speed Of Light.

The concept of an infinite quantity of layers of Dimension's, Planes and Timelines all on top of each other is not absurd: Why does Death not exist for Life, why do we no longer Sensai The Dead, which sounds like a B—Stupid Violent Black Horror Film but now in the Light and Shadow of these arguments, evidences and proofs makes perfect sense.

No, the only form of extra speed of light travel which makes sense is Near-Insta or Insta Teleportation through a Doorway, Door, Portal and/or Gate by transfer of the massless Information through the Energy Matrix of Reality itself. By definition, if it does not have mass, than it can travel faster than Light Speed.

However, now with no problem with doing Time anymore, and Shadow Speed, we don't have to worry about mass anymore since there is no more resistance factor.

This then also deletes Time out of the Equation, no longer important, insignificant. Our

Universe is now open to Human exploration! Famous last words . . . If you've ever played Dungeons & Dragons, 1st or 2nd Edition, then you know what I mean . . .

And here again, just to keep the objectivity, and punch holes in my own arguments, too, not to mention those of Stargate, again, you still need a Receivor Transfer Transform Interface Device, which somehow miraculously converts, without error down to the last genome, the Information back into Energy Matter, considering various OS's to date and all the hacking going on with Reseller Galaxy I don't think it's going to be an easy ride.

We're also still left with the problem of how to get there and set up Relay-Router Space Station's. And the time it takes to set up 01 Mining Colony Planet has now made it absurd, how are we supposed to do any of these things before even 2103??

It's actually futile to argue Spirit and Consciousness these are in-and-out-of-body-travel-and-dreams-experiences. For whereas, highly probable, being massless, if I can only get there as a Ghost of Humanity then we will never colonize our Universe.

It also take a stupendous quantity and quality of Work to populate the Solar System.

Is Time only relative? Within Absolute Infinity, being timeless, such relativity does not exist. Also, there is no definable limit at how fast any object can turn or move. It's almost like, again, does Time = 0? See previous chapter here for more details on such.

Is Time a particle? Does it bind itself to Matter? Then, once again, it only equals Particle Motion's which are unclarified and uncertain. How fast something turns, there being no actual limit in Infinity of various Virtual Object's, is purely relative, subjective, qualititative, a virtual value which Human again applied to the objects, which can be measured to Finite, Near-Infinite and Infinite. There are infinitesimal quantities of points in one line, spiral, circle, ellipse or sphere. Such now becomes only fluidic Space.

For each moment there is still yet another smaller moment. It remains Motion only.

The only possible explanation, by reduction, is there is some kind of inherent Natural Mechanism in Reality itself whichs allows you to aim at some point in Deep Space and once your collective Information reaches there you rematierialize into the Energy Matrix, like a Child from his/her Mother's womb. Maybe you can carry the program and device with you for such is also Information to Energy to Matter with your Soul to Spirit to Mind to Body to all your other permutations as your waves and particles recoalesce.

And finally, the stunning conclusion, Time cannot exist in Infinity, for by definition, there is no Time in Infinity. If, however, our Universe is only Near-Infinite or Finite, denying GOD, God's and Goddesses, Immortality, Heroes and Villain's then Time can only exist as a purely qualitative relative value which we arbitrarily apply to an object.

Evolutionary Essays—End of Part 01

Kyle Lance Proudfoot

I first started in IT in 1997 when I started my website Silverlingo.com, but I was actually already busy since arcades in Toronto, Canada. From 2001 to 2005 I worked as a System Administrator in diverse instances, since then as a Webmaster, Webdesigner and/or Webdeveloper on about 20 websites. I have also done projects such as writing, editing and translating text including publishing 5 books, made music, art and B- humor film plus Open Source scripting and 2D/3D graphic design. My most recent project which is also continuous is my Photography and Video which I am busy uploading to Google+.

My passion is writing in genres such as Philosophy, Science Fiction, Fantasy, Humor, Poetry, Politic's, Psychology and now even in Sciences. See also my sister website Planesofexistence.eu which is dedicated to my writing.

Next to IT I have a lot of knowledge and work experience in Literature, Music, Art, Film, 3D Games, Internet, Multimedia, Entertainment and Social Media to develop projects independently or with others to make Open Source to Retail products. So far over 3 decades I have done 30000+ sessions of The Free Show, watched 9000+ Film's and played most 3D Games except the most recent ones.

My goals for the future are to make Literature, Film, Music, Photography, Art, Video, Audio, Science Fiction, Fantasy, Humor, Poetry, Philosophy, Psychology, Politic's, Space Travel, The Free Show, Biking, Nature, 3D Games, Cooking.